JUST-IN-TIME SCHEDULING:

Models and Algorithms for Computer and Manufacturing Systems

Recent titles in the
INTERNATIONAL SERIES IN OPERATIONS RESEARCH & MANAGEMENT SCIENCE
Frederick S. Hillier, Series Editor, *Stanford University*

Gass & Assad/ *AN ANNOTATED TIMELINE OF OPERATIONS RESEARCH: An Informal History*

Greenberg/ *TUTORIALS ON EMERGING METHODOLOGIES AND APPLICATIONS IN OPERATIONS RESEARCH*

Weber/ *UNCERTAINTY IN THE ELECTRIC POWER INDUSTRY: Methods and Models for Decision Support*

Figueira, Greco & Ehrgott/ *MULTIPLE CRITERIA DECISION ANALYSIS: State of the Art Surveys*

Reveliotis/ *REAL-TIME MANAGEMENT OF RESOURCE ALLOCATIONS SYSTEMS: A Discrete Event Systems Approach*

Kall & Mayer/ *STOCHASTIC LINEAR PROGRAMMING: Models, Theory, and Computation*

Sethi, Yan & Zhang/ *INVENTORY AND SUPPLY CHAIN MANAGEMENT WITH FORECAST UPDATES*

Cox/ *QUANTITATIVE HEALTH RISK ANALYSIS METHODS: Modeling the Human Health Impacts of Antibiotics Used in Food Animals*

Ching & Ng/*MARKOV CHAINS: Models, Algorithms and Applications*

Li & Sun/ *NONLINEAR INTEGER PROGRAMMING*

Kaliszewski/ *SOFT COMPUTING FOR COMPLEX MULTIPLE CRITERIA DECISION MAKING*

Bouyssou et al/ *EVALUATION AND DECISION MODELS WITH MULTIPLE CRITERIA: Stepping stones for the analyst*

Blecker & Friedrich/ *MASS CUSTOMIZATION: Challenges and Solutions*

Appa, Pitsoulis & Williams/ *HANDBOOK ON MODELLING FOR DISCRETE OPTIMIZATION*

Herrmann/ *HANDBOOK OF PRODUCTION SCHEDULING*

Axsäter/ *INVENTORY CONTROL, 2nd Ed.*

Hall/ *PATIENT FLOW: Reducing Delay in Healthcare Delivery*

Józefowska & Węglarz/ *PERSPECTIVES IN MODERN PROJECT SCHEDULING*

Tian & Zhang/ *VACATION QUEUEING MODELS: Theory and Applications*

Yan, Yin & Zhang/ *STOCHASTIC PROCESSES, OPTIMIZATION, AND CONTROL THEORY APPLICATIONS IN FINANCIAL ENGINEERING, QUEUEING NETWORKS, AND MANUFACTURING SYSTEMS*

Saaty & Vargas/ *DECISION MAKING WITH THE ANALYTIC NETWORK PROCESS: Economic, Political, Social & Technological Applications w. Benefits, Opportunities, Costs & Risks*

Yu/ *TECHNOLOGY PORTFOLIO PLANNING AND MANAGEMENT: Practical Concepts and Tools*

Kandiller/ *PRINCIPLES OF MATHEMATICS IN OPERATIONS RESEARCH*

Lee & Lee/ *BUILDING SUPPLY CHAIN EXCELLENCE IN EMERGING ECONOMIES*

Weintraub/ *MANAGEMENT OF NATURAL RESOURCES: A Handbook of Operations Research Models, Algorithms, and Implementations*

Hooker/ *INTEGRATED METHODS FOR OPTIMIZATION*

Dawande et al/ *THROUGHPUT OPTIMIZATION IN ROBOTIC CELLS*

Friesz/ *NETWORK SCIENCE, NONLINEAR SCIENCE AND DYNAMIC GAME THEORY APPLIED TO THE STUDY OF INFRASTRUCTURE SYSTEMS*

Cai, Sha & Wong/ *TIME-VARYING NETWORK OPTIMIZATION*

Mamon & Elliott/ *HIDDEN MARKOV MODELS IN FINANCE*

del Castillo/ *PROCESS OPTIMIZATION: A Statistical Approach*

**** A list of the early publications in the series is at the end of the book ****

JUST-IN-TIME SCHEDULING:

Models and Algorithms for Computer and Manufacturing Systems

Edited by

Joanna Józefowska

Poznań University of Technology
Poznań, Poland

Springer

Joanna Józefowska
Poznań University of Technology
Poznań, Poland

Series Editor
Fred Hillier
Standford University
Standford, CA, USA

Library of Congress CContract Number : 2007926587

ISBN 978-387-71717-3 ISBN 978-0-387-71718-0 (e-book)

Printed on acid-free paper.

9 8 7 6 5 4 3 2 1

springer.com

To my mother

Preface

The philosophy of just-in-time manufacturing was first introduced by the Japanese automobile producer Toyota in 1950s. This philosophy may be briefly defined as elimination of waste and simultaneous continuous improvement of productivity. There are many different sources of waste in a manufacturing system; therefore, many activities need to be undertaken in a company in order to effectively implement the just-in-time philosophy. Waiting time, overproduction and inventory are the sources of waste which can be eliminated by appropriate production planning and scheduling.

The goals of just-in-time scheduling differ from the goals considered in traditional production scheduling. Therefore new scheduling problems have been defined within the theory of scheduling to meet the need for practical solutions. Two optimization objectives are considered in the context of just-in-time scheduling. The first one is minimization of production variation, which means that the same amount of any output should be produced every day, or even every hour. An interesting aspect of the problem of minimizing the production variation is its similarity to the problem of apportionment. The results of the theory of apportionment are exploited in the analysis and design of scheduling algorithms used to minimize the production variation. The second objective examined in just-in-time scheduling is minimizing the total earliness and tardiness cost. Minimization of the total earliness and tardiness cost expresses the aim to reduce inventory cost and, simultaneously, satisfy the customer demands with timely delivery of products. This objective gives rise to non-regular performance measures, and thus leads to new methodological issues in the design of scheduling algorithms.

Scheduling problems with both objective functions, i.e. minimization of the production variation and minimization of the earliness and tardiness cost, which appear in just-in-time production planning and control systems, have found numerous applications in the control of computer systems. The most important class of computer systems working in a just-in-time environment is the class of the real-time systems. The main requirement for a real-time system is to respond to externally generated input stimuli within a finite and specified period. This requirement results in the same scheduling objectives as those considered in the just-in-time manufacturing systems. Consequently, the same optimization algorithms may be applied to solve scheduling problems in just-in-time manufacturing systems and in real-time computer systems.

The aim of this book is to present both classes of scheduling problems and both application areas together, in order to show the similarities and differences of the approaches. The book contains a survey of exact and heuristic algorithms developed to solve the scheduling problems in the just-in-time environment. The presented survey may alert the reader to similarities of models and techniques used in different optimization domains, like those for the scheduling theory and the apportionment theory. Many concepts and algorithms are illustrated with examples, tables and figures to enhance the clarity of the presentation. As such, this book should differ from other surveys of just-in-time scheduling problems in the scope and the unified treatment of problem formulation and solution procedures.

The intended audience of this book includes professionals, researchers, PhD students and graduate students in the fields of Operations Research & Management, Business Administration, Industrial Engineering, Applied Mathematics, System Analysis as well as Computer Science and Engineering.

The book is divided into five chapters. In Chapter 1 a brief presentation of the application areas of the considered scheduling problems is provided. First, the principles of the just-in-time production planning and scheduling are discussed. Second, the basic features of the real-time systems are characterized. The basic terminology is introduced and the motivation for the research presented in the following chapters is given. Chapter 2 contains an introduction to two optimization domains, the theory of scheduling and the theory of apportionment. Results obtained within these two theories are used to solving the just-in-time scheduling problems discussed later. Chapters 3 and 4 present the problems and algorithms for minimizing the earliness/tardiness cost. Chapter 3 focuses on the case of the common due date and Chapter 4 on the case of task dependent due dates. Finally, in Chapter 5 the problems and algorithms for minimizing the production variation are examined. Within this approach some real-time scheduling problems are discussed.

Despite great effort involved in the preparation of this book, the author is aware that avoiding all errors is impossible. Taking all responsibility for possible deficiencies, the author would welcome any comments on the book.

The author would like to express her thanks for all the help and encouragement received during the preparation of this book from colleagues, friends and family.

Poznań–Cottbus, March 2007 *Joanna Józefowska*

Contents

1

Just-in-time concept in manufacturing and computer systems

The scheduling problems presented in this book follow from two different application areas: production and computer engineering. The common feature of the considered problems is the fact that it is crucial for the control system to observe the due dates.

Just-in-time is a production planning and control philosophy that seeks to eliminate waste. Completing a task before or after the due date incurs additional cost, which is a waste of resources. Therefore in just-in-time systems it is desirable to complete a task as close to its due date as possible.

In computer systems the idea of observing due dates is typical for real-time systems. A system is said to be real-time if the correctness of an operation depends not only upon the logical correctness of the operation but also upon the time at which it is performed.

It appears that the same classes of scheduling problems and algorithms are applied in the control of just-in-time production systems as well as real-time computer systems.

In this chapter we introduce the basic ideas characterizing the just-in-time and real-time systems. The aim is to provide a brief survey of aspects that are relevant to the main topic of this book.

1.1 Manufacturing systems

Although a quite recent concept, jut-in-time (JIT) is one of the fundamental approaches to modern manufacturing planning and control. Just-in-time represents a philosophy, as well as a set of techniques developed for production planning and control.

1.1.1 Production planning and control

Production planning and control (PPC) is responsible for the planning and control of the flow of materials through the manufacturing process. The inputs to the production planning system are:

- *product structure* given by engineering drawings and specifications or by a *bill of material (BOM)*;
- *process specifications* which describe the steps necessary to make the finished product, i.e. sequence of operations required to make the product, equipment and accessories required, and standard time required to perform each operation;
- *available facilities* including work centers, equipment and labor;
- *quantities required*, this information comes usually from forecast, customer orders and the material requirements plan.

According to the APICS Dictionary [76], every part or assembly in a product structure is assigned a level code signifying the relative level in which that part or assembly is used within the product structure. The same dictionary defines the *bill of material (BOM)* as a listing of all the subassemblies, intermediates, parts and raw materials that go into a parent assembly, showing the quantity of each required to make an assembly. Further on we will use the term *finished product* or *product* for an end item, i.e. item produced at level 1 - the final assembly line. An item completed at any level (product, subassembly, part, material) will be called an *output.* Calculation of the output demands of each finished product, following from the BOM, is called the *BOM explosion* .

A *work center* is a specific production area, consisting of one or more machines with identical capabilities that can be considered as one unit for purposes of capacity requirements planning and detailed scheduling. In scheduling, the term machine is used for simplicity.

The production of goods is often executed in batches. A *batch* is the number of units made between sequencial setups at a work center. *Setup* is the work required to change a specific work center from making the last good piece of product A to the first good piece of product B.

An important process parameter is the *lead time*, which is determined as the time span needed to perform the process. It includes order preparation, queuing, processing, moving, receiving and inspecting, and any expected delays.

The availability of a facility is measured by its *capacity* which is the quantity of work that can be performed by the facility in a given period.

A market oriented company focuses on meeting the customer expectations with regard to product quality and delivery lead time. Four basic manufacturing strategies are distinguished with regard to the product design, manufacturing and inventory stages included in the delivery lead time. These are: *engineer-to-order, make-to-order, assemble-to-order* and *make-to-stock*. The delivery lead time is longest for the engineer-to-order and shortest for the make-to-stock strategy. Engineer-to-order means that the delivery lead time encompasses the production process from the product design phase until the final shipment. No materials are purchased until needed by manufacturing. Make-to-order means that the manufacturer does not start to make the product until a customer order is received. Some inventory of raw material is maintained. Assemble-to-order means that the product is made from standard components that are kept in the inventory and assembled according to the customer order. Make-to-stock means that the supplier manufactures the goods and sells them from the finished product inventory.

The main goal of production planning and control is devising plans to balance the demands of the marketplace with the resources and capacity of the manufacturing system. This goal is realized hierarchically, from long range to short range, at the following five levels [10]:

- strategic business planning,
- sales and operations planning (SOP),
- master production scheduling (MPS),
- material requirements planning (MRP),
- purchasing and production activity control.

Each level varies in purpose, time span, and level of detail. As we move from strategic plan to production activity control, the purpose changes from general direction to specific detailed planning, the time span decreases from years to days, and the level of detail increases from general categories to individual components and machines.

The strategic business plan is a statement of the major objectives the company expects to achieve over the next two to ten years. This plan is based on the long-range forecast and includes participation from marketing, finance, production and engineering.

Given the objectives, set by the strategic plan, the sales and production plan is created. The production management determines the quantities of each product group that must be produced in each period, the desired inventory level, the resources of equipment and material needed in each period and the availability of the resources needed. One of the most important inventories is the *work in process*. Work in process is a product or products in various stages of completion throughout the manufacturing process, including all material from raw material that has been released for initial processing up to completely processed material awaiting final inspection and acceptance as a finished product. The *planning horizon*, i.e. the time span from now to some time in the future for which the plan is created, for the sales and operations planning is usually six to eighteen month.

The master production schedule is a plan for the production of finished products. It breaks down the sales and operations plan to show, for each period, the quantity of each finished product to be made. The planning horizon for the master production plan usually extends from three to eighteen months and depends on the purchasing and manufacturing lead times.

The objective of material requirements planning is to create a plan for production and purchase of the components used in making the items in the master production schedule. MRP begins with the items listed on the master production schedule and determines:

(i) the quantity of all components and materials required to fabricate those items,
(ii) the due date that the components and materials are required.

The plan is accomplished by exploding the bill of material, adjusting for inventory quantities on hand or on order, and offsetting the net requirements by the appropriate lead times. The level of detail is high and the planning horizon is at least as long as the combined purchase and manufacturing lead times.

Purchasing and production activity control represent the implementation and control phase of the production planning and control system. Purchasing is responsible for establishing and controlling the flow of raw materials into the factory. Production activity control is responsible for planning and controlling the flow of work through the factory. The level of detail is high and plans are reviewed and revised daily.

Two types of production systems may be distinguished: *push* and *pull* systems. Push systems are typically centralized systems, where all forecasting and order decisions are made centrally. It results in production of items at times required by a given schedule planned in advance. The pull systems are typically decentralized. In such systems items are produced only as demanded for use or to replace those taken for use.

An integrated method for planning all resources of a manufacturing company is called *manufacturing resource planning (MRP II)*.

To get the most profit a company must aim at providing the best customer service, with minimimum production and distribution costs, and minimimum inventories. Usually, providing good customer service conflicts with minimization of production costs and inventory.

Production planning and control system that explicitly sets the goal to supply the customers with what they want when they want it and to keep inventories at a minimum is called a just-in-time system.

1.1.2 Just-in-time systems

There are many different definitions of just-in-time concept and they still evolve to become more general. An intuitive, although not precise description says that just-in-time philosophy pursues zero inventories, zero transactions and zero disturbances, the last one understood as a routine execution of schedules.

More precisely, the just-in-time approach aims at reducing ([247]):

- the complexity of the detailed material planning,
- the need for shop-floor control,
- the desired inventory levels,
- the transactions associated with shop-floor and purchasing systems.

The achievement of JIT goals requires deep changes in the manufacturing process. One of such changes is reduction of setup times, followed by reduction of batch sizes. This complies with reduction of inventories. Setup time may be reduced by applying new technologies, like CNC (Computer Numerical Control) machines, or by thorough analysis of the setup process leading to better execution of the process. The results are sometimes surprising. It has been observed that setup time can decrease from several hours to less than ten minutes.

Another important requirement is constant improvement of production quality through improvement of the manufacturing process. The reason for this requirement is that any quality problem may result in a stoppage of the entire manufacturing system, unless undesirable buffer inventories are held. The survey of various methods of quality improvement is beyond the scope of this book. Let us just mention two, the preventive maintenance and foolproof operations which are typical for the total quality management applied in just-in-time systems. Preventive maintenance expresses the necessity to pay the same attention to product quality as to equipment and process quality. The foolproof operations are checking operations built into the manufacturing processes so that the quality of a part is evaluated as the part is processed. In consequence, any defects are found immediately as they occur, decreasing the cost or removal of repair of the faulty part.

JIT program includes continuous improvement as the maxim for day-to-day operation. Every worker is supposed to get better in some dimension, such as fewer defects, more output, or fewer stoppages. Thousands of small improvements in methods, processes and products build a continuous struggle for excellence. Every worker should be deeply involved and actively participate in this process.

The most common organizational structure in the JIT environment is a *production cell.* A production cell is a group of machines manufacturing a particular set of parts. The layout of the machines should minimize transportation distance and inventories between consecutive operations. Cells are usually U-shaped to increase worker interaction and reduce material handling. Each worker is trained to operate several different machines to increase flexibility of the system in handling changes in production mix or volume.

The most important changes resulting from applying the JIT philosophy are observed in the manufacturing planning and control(MPC) system. Although JIT plays the main role in production planning and control, it affects also:

- process design,
- product design,
- human/organizational elements.

The main implications of JIT for the process design include high quality of the manufacturing process, design for manufacturing in cells and reducing the number of levels in the bill of material. It is also worth mentioning that some product redesign may lead to part unification meaning that the same parts are used in assembly of various finished products. Such solution also reduces the volume on inventories held.

Appropriate product design opens opportunities for process improvement. The goals of the process improvement are the reduction of the number of setups and more flexible use of the equipment, allowing easy adjustment to the changing production demand. Moreover, if BOM is reduced to two or three levels, the cost of performing the detailed planning is cut significantly.

Human/organizational elements are also involved in building the JIT philosophy. The workers are the most valuable asset of the company. The company continuously invests in education and training of the employees. The employees are capable of performing various task, not only machining operations. They are involved in the process organization including scheduling, equipment maintenance or data entry in the cell. They are expected to use their skills to constantly improve the process.

The idea of just-in-time production control originated in Japan. An approach aiming at reduction of the inventory cost was first introduced in Toyota. The just-in-time production system applied in Toyota is described in [201]. The main idea was to complete the orders as close as possible to the desired due dates. The main difficulties were to adjust the production system in such a way that additional organization cost did not exceed the savings in inventory cost.

The way JIT is implemented in Toyota is called a kanban system. The kanban system works as follows. The items (raw materials, parts, subassemblies and assemblies) are stored and transported in boxes, with a so called, kanban card attached to each box. Of course, only items of the same type may be placed in the same box. Assume we have a production system consisting of three production cells and three assembly cells, as illustrated in Figure 1.1. If an item is needed in the assembly cell, the appropriate box is taken from the storage area and the accompanying kanban is removed from the box and placed on a special board. The kanban cards on the board represent the production orders for the relevant production cell. The number of kanban cards in the system limits the inventory since production of any part will not start unless the kanban is released and placed on the board.

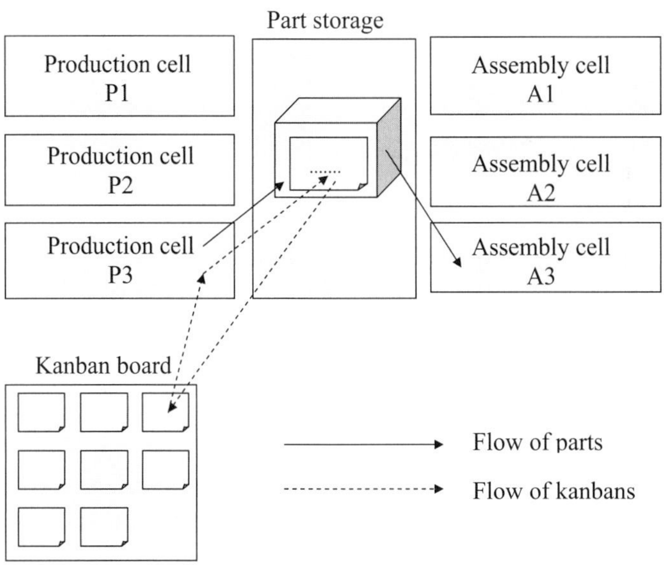

Fig. 1.1. A single kanban system.

In fact, the Toyota kanban system is more complicated than the single kanban system illustrated in Figure 1.1. Since the storage area after production is separated from the storage area before assembly, there are two types of kanban cards used in this system conveyance cards and production cards. Conveyance cards are used to move containers from one storage area to another. Production cards authorize the production of parts etc. The kanban system can be extended back to the suppliers. Some of the Toyota suppliers are included in this system.

The kanban system is a pull system since any activity can start only if authorized by a relevant kanban card. No work center is allowed to produce or transport any part just in order to keep workers busy.

The number of kanban cards handled in the system determines the amount of work in process inventory. An optimal number of kanban cards is calculated depending on the demand per unit of time, the lead time, the safety stock and the container capacity. The amount of the safety stock should be kept small, not exceeding 10 percent of the demand per time unit. The containers are kept small and standard. At Toyota the container does not hold more than 10 percent of daily demand.

According to [247], the following rules have to be observed in order to keep the system operate correctly:

- an appropriate kanban card must be attached to each container;
- containers have standard capacity and only a full container may be transferred;
- only the production cell using the items may initialize the transfer of parts, never the provider;
- parts may be removed from storage only if an appropriate kanban card is received;
- production of parts may be launched only upon receipt of an appropriate kanban card.

Following these rules, combined with continuous struggle for improvement, i.e. reduction of the number of kanban cards, is the way Toyota realizes the JIT production. The Toyota planning system is organized as follows. At the higher level of the production planning system a great attention is paid to keep stable production rates. The planning horizon covers one year and is updated once a month. The master production schedule is the final assembly schedule. A detailed daily schedule is known for three months ahead and is identical throughout this period. High production stability is crucial for the application of the kanban control system. Constructing a master production schedule that maintains the stability of production at all levels is a complex scheduling task. It gave rise to scheduling problems called Output Rate

Variation (ORV) problem and Product Rate Variation (PRV) problem that are illustrated with some numerical examples in Section 1.1.3. Mathematical formulation and scheduling algorithms for the PRV and ORV problems are presented in Chapter 5.

A different approach has to be taken when the organizational requirements for balanced production cannot be met. Then the main requirement is to complete the items as close to their due dates as possible. Problems of this type are described in Section 1.1.4.

Although the main goal of JIT is to reduce inventories, the way it is implemented offers many more benefits (see [97, 109, 247]), including:

- better product quality,
- shorter delivery times,
- better flexibility in terms of production mix and volume,
- more friendly working environment: the worker performs various tasks and takes part in the management decisions at the shop-floor level,
- more efficient utilization of workers capabilities,
- good relationships with suppliers, which result in increased reliability of supplies.

Concluding, there are significant benefits from just-in-time production planning and control; however, the system requires high organization discipline. Moreover, it is less risky to apply JIT in manufacturing systems with repetitive production and simple product structures.

An extensive literature review concerning the just-in-time philosophy is presented in [105]. Optimization models of kanban based production control are discussed e.g. in [31, 208, 114]. Mathematical programming models are formulated for different optimality criteria including the inventory cost, shortage cost and setup cost.

1.1.3 Balanced schedules

As we mentioned in the previous section, a method to reduce the scheduling cost in a just-in-time system is to organize a very stable production environment at all levels. Stable production means that the schedule of a production cell is fixed and repeated every day, i.e. parts are produced in the same order and quantities day by day. Such assumption, however, imposes strong constraints on the production schedule at the final assembly line level. In order to keep the inventories low, the assembly line should use the same quantities of components every day, or even every hour. In general, different finished products require different components, so keeping the part usage constant, and, at the same time meeting the customer requirements is difficult.

The just-in-time sequencing has proven to be a universal and robust tool used to balance workloads throughout just-in-time supply chains intended for low-volume high-mix family of products, Kubiak [160]. It renders supply chains more stable and carrying less inventories of finished products and components but at the same time it ensures less shortages. The original kanban system that was the first implementation of the just-in-time philosophy did not assume any explicitly defined due dates.

Since JIT is a pull system, the production orders are initialized at the final assembly line. The master production schedule (MPS) determines the required quantities and due dates of individual finished products. Using the bill of material the quantities of raw materials, parts, and subassemblies needed to complete the products determined by the MPS are calculated. Thus the demands for subassemblies and parts at the remaining levels follow directly from the MPS and BOM.

The following example is developed to illustrate how the MPS and BOM are used to calculate the demands for outputs at all production levels. Further, we use the same example to explain the idea of a balanced schedule.

Example 1.1. Figure 1.2 shows product structure which is identical for four products: A, B, C and D. The corresponding output demands of each finished product are presented in Table 1.1, and the weekly product demands in Table 1.2. The resulting weekly demand for particular outputs is given in Table 1.3.

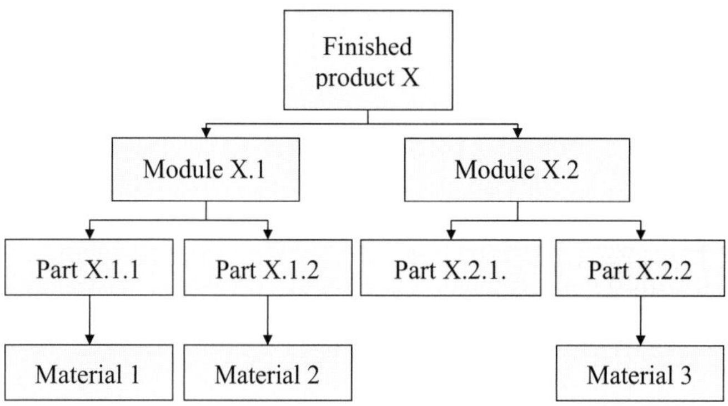

Fig. 1.2. Product structure.

Table 1.1. Output demands of each finished product

Components	Product A	Product B	Product C	Product D
Module X.1.a	1	0	0	1
Module X.1.b	0	1	1	0
Module X.2.a	1	0	1	0
Module X.2.b	0	1	0	1
Part X.1.1.a	2	0	0	1
Part X.1.1.b	0	1	3	0
Part X.1.2.a	0	1	0	2
Part X.1.2.b	1	0	2	0
Part X.2.1.a	3	2	0	0
Part X.2.1.b	0	0	2	3
Part X.2.2.a	3	0	0	0
Part X.2.2.b	0	3	0	2
Part X.2.2.c	0	0	3	0
Material 1	1	5	1	4
Material 2	4	1	3	1
Material 3	2	2	2	2

Table 1.2. Weekly product demands

	Product A	Product B	Product C	Product D
Weekly demand (in units)	40	160	80	120

Another data that has to be taken into account are the process specifications of particular parts. The parts are fabricated in two production cells. The first cell consists of 7 machines, and the second one of 8 machines. The process specifications of the parts are presented in Tables 1.4 and 1.5. We assume that the final assembly line operates 5 days a week, 8 hours a day and has the capacity to produce 10 units of any product in an hour.

There are two types [188] of master production schedules: *a batch schedule* and *a balanced schedule.* A batch schedule aims at minimization of the number of setups in the final assembly line. Therefore the total demand for each product in the planning horizon is produced in a single batch. It is easy to notice, however, that this approach results in large inventories of the finished products because the production in the final assembly line is not synchronized with the customer demand. Moreover, at the part production level, the demand for components is

Table 1.3. Weekly output demands

Item	Demand				
	Product A	Product B	Product C	Product D	Weekly total
Module X.1.a	40	0	0	120	160
Module X.1.b	0	160	80	0	240
Module X.2.a	0	0	80	0	120
Module X.2.b	40	160	0	120	280
Part X.1.1.a	80	0	0	120	200
Part X.1.1.b	0	160	240	0	400
Part X.1.2.a	0	160	0	240	400
Part X.1.2.b	40	0	160	0	200
Part X.2.1.a	120	320	0	0	440
Part X.2.1.b	0	0	160	360	520
Part X.2.2.a	120	0	0	0	120
Part X.2.2.b	0	480	0	240	720
Part X.2.2.c	0	0	240	0	240
Material 1	0	800	0	480	1400
Material 2	160	0	240	0	690
Material 3	80	320	160	240	800

Table 1.4. Process specifications of the parts produced in Cell 1

Item	Machine						
	1	2	3	4	5	6	7
Part X.1.1.a	1	1	1	1	0	1	1
Part X.1.2.b	1	1	0	1	1	1	1
Part X.2.1.a	1	0	1	1	1	1	1
Part X.2.2.b	0	1	1	0	1	0	0
Part X.2.2.c	1	1	0	1	0	1	1

Table 1.5. Process specifications of the parts produced in Cell 2

Item	Machine							
	8	9	10	11	12	13	14	15
Part X.1.1.b	1	1	1	1	1	0	1	1
Part X.1.2.a	1	1	1	1	0	1	1	0
Part X.2.1.b	1	1	1	1	1	1	0	1
Part X.2.2.a	0	1	1	1	1	0	1	1

high during the period of production and very low when the product is not assembled. Thus the production of components is not stable.

A balanced schedule aims at better synchronization of the output with the market demands. It means smaller batches with completion times more equally distributed over the planning horizon. Consequently, inventories are kept at a low level and production of components is more stable. Obviously, the number of setups increases. Thus setup times and setup cost should be minimized already during product design and planning. Continuing Example 1.1, we will justify the use of balanced schedules in just-in-time systems.

A batch schedule for Example 1.1 is constructed as follows. The total demand for Product A is scheduled on Monday, together with 40 units of Product B. Production of B is continued on Tuesday (80 units) and Wednesday (40 units). Another 40 units of Product C are scheduled on Wednesday. On Thursday the remaining 40 units of product C and 40 units of Product D are produced. Finally, on Friday the remaining 80 units of Product D are completed.

A balanced schedule consists of five identical schedules for each day from Monday through Friday. Each day 16 units of Product A, 30 units of Product B, 14 units of Product C, and 20 units of Product D are made.

Both schedules are presented in Figure 1.3.

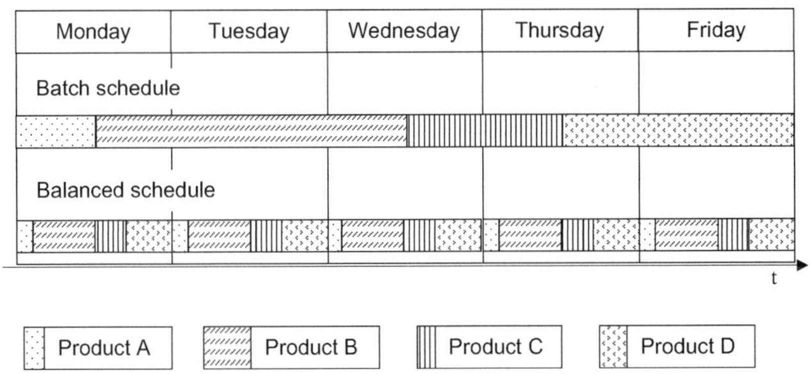

Fig. 1.3. Batch and balanced schedules.

Two observations follow immediately from the comparison of both schedules. The first one is that there are only four assembly line setups in the batch schedule, while in the balanced schedule there are four setups daily. This justifies the requirement to minimize the setup time in just-in-time systems. The second observation is that the part usage, as well as production cell workload, are much more even when the balanced schedule is applied. The part usage on particular days for both schedules is presented in Table 1.6 and the required cell capacity in Table 1.7.

Table 1.6. Part usage on particular days

	Batch schedule					Balanced schedule
Part	Monday	Tuesday	Wednesday	Thursday	Friday	Daily
X.1.1.a	80	0	0	40	80	40
X.1.1.b	40	80	160	120	0	80
X.1.2.a	40	80	40	80	160	80
X.1.2.b	40	0	80	80	0	40
X.2.1.a	200	160	80	0	0	88
X.2.1.b	0	0	80	200	240	104
X.2.2.a	120	0	0	0	0	24
X.2.2.b	120	240	120	80	160	144
X.2.2.c	0	0	120	120	0	48

The variations in part usage on consecutive days are significant. For example, part X.2.1.a is used only from Monday through Wednesday, while part X.2.1.b from Wednesday through Friday. The maximum demand for part X.2.1.b is 240 units (on Friday) and minimum demand is zero (on Monday and Tuesday). In order to meet such requirements, the schedules at the production cells must be very complicated and inventories high. In the balanced schedule the daily production of every part is equal, making the schedule more stable and creating less inventories.

Table 1.7. Required cell capacity on particular days

	Batch schedule					Balanced schedule
	Monday	Tuesday	Wednesday	Thursday	Friday	Daily
Cell 1	2280	1680	1920	1560	960	1680
Cell 2	2040	2400	2040	2360	3040	2376

The capacity requirements in Cell 1 vary from 960 units on Friday to 2280 units on Monday. It means that the required capacity in Cell 1 is more than two times bigger on Monday than on Friday. The variation of capacity requirements in Cell 2 is not so big but it is also significant. Such differences lead to low machine utilization. An illustration of the variation of machine utilization calculated for Machine 4 in Cell 1 on particular days is shown in Figure 1.4. Large inventories of finished products may be also expected since some items of Product A may wait until the end of the week before being shipped to the customer.

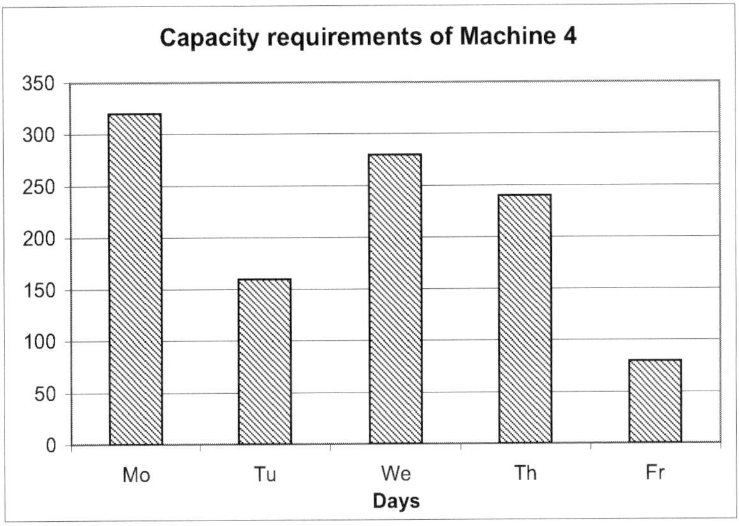

Fig. 1.4. Utilization of Machine 4 in the batch schedule.

In the balanced schedule proposed in Figure 1.3 the part usage as well as the machine and cell utilization is equal each day of the week. The benefits of such solution are the following [188]:

- it is easy to schedule the manpower,
- production at all levels is more routine,
- it is easier to develop procedures to improve quality,
- less inventory is carried,
- the product is shipped to the customer faster.

Closer examination of the balanced schedule in Figure 1.3 leads to an observation that a more balanced schedule may be achieved, for example, by repeating the sequence of products BDCBADBCDB every hour. In such schedule the part usage, as well as the machine utilization, are identical every hour, not only every day.

The above considerations lead to the formulation of the optimization problem called Output Rate Variation Problem. The same problem reduced to a single level problem, i.e. concentrating on the assembly line schedule only, is called Product Rate Variation Problem. These problems will be formulated in Chapter 5.

1.1.4 Earliness and tardiness cost

As we mentioned at the beginning of Section 1.1.2, a JIT system, especially in the form implemented in Toyota, requires high organization discipline. Moreover, its implementation is easier in manufacturing systems with repetitive production and simple product structures. Very often it is difficult to meet these requirements. Still, the reduction of inventory cost remains a serious challenge. If the customer demands are known at the beginning of the planning horizon, it is natural to define due dates of particular orders. Starting from the master production schedule, where the due dates of finished products are determined, the information is propagated down to the remaining production levels in order to calculate due dates for subassemblies, parts and raw materials.

Obviously, completing a production order after its due date is undesirable. In case of finished products, a late completion leads not only to customer dissatisfaction, but may result even in order or customer loss. If a production order is late for assembly, a stoppage at the assembly line may occur. In general, cost incurred by late completion of an order include invoking contractual penalty clauses, cost of dealing with the customer, cost of expediting tardy tasks and cost of updating the production schedule.

On the other hand, outputs completed before their due dates have to be stored and create undesired inventory. Inventory cost depend on inventory quantity, item's value and length of time the inventory is carried. The inventory costs include the cost of capital committed to the inventory, cost of insurance taken on inventories, cost of inventory obsolescence or product shelf life limitations and operating cost involved in storing the inventory. The cost of capital is usually expressed as an annual interest rate based on the cost of obtaining bank loans to finance the inventory investment. Operating costs include the rental cost of public warehousing space, or cost of owning and operating warehouse facilities (such as heating, light and labor).

In traditional production planning and control systems based on manufacturing resource planning (MRP II), the considered scheduling objectives are minimization of product flow time, maximization of machine utilization and avoidance of late task completion. Just-in-time approach introduced in Japanese factories drew the attention to minimization of inventory cost. As a result a new class of scheduling problems emerged with the optimality criteria defined as total earliness and tardiness cost.

The vast majority of research is devoted to single machine systems, although parallel and dedicated machine systems are also examined. In many practical situations it is justified to consider the production cell, or final assembly line as a single machine, since the sequence of machining operations is fully automated and from the scheduling perspective may be considered as a single task.

In some situations all tasks may share the same due date. Scheduling problems with a common due date have some properties which make the scheduling decisions easier. Assuming a common due date has also strong practical rationale. An example may be a production system supplying an assembly line where numerous parts are required at the assembly line at the same moment, when the assembly process starts. Another situation where a common due date is justified occurs when finished products are shipped to the customer. In order to better utilize the means of transportation all products to be shipped together should be completed at the same time.

An interesting aspect of just-in-time scheduling is the possibility of due date negotiation. The due date then becomes a decision variable. The practical sense of due date assignment becomes clear when a company offers a price reduction to its customer if the due date is set later than expected. Distant due date gives more chance to complete the order in time and project the right image of the producer. Sometimes a reasonable inventory may be accepted if it helps to meet the due dates fixed in the contract.

Another modification of the just-in-time scheduling problem that is often considered is the possibility to reduce the task processing time to meet the required due dates. Reducing the task processing time incurs additional cost, for example more resources need to be used. This cost is added to the earliness and tardiness cost in the objective function, and the goal is to minimize the total cost.

The scheduling problems with minimization of the total earliness/tardiness cost occur in other areas, not only in just-in-time production systems. Many activities such as purchasing, transportation, maintenance or military operations require just-in-time control. One of the most important areas where this approach leads to significant cost reduction is logistics. Logistics is the technique of managing and controlling the flow of goods, energy, information and other resources like products, services, and people, from the source of production to the marketplace. In simple words, logistics can be defined as having the right quantity at the right time for the right price. It follows from this definition that just-in-time approach realizes the logistics objectives.

Just-in-time philosophy is also applied in computer systems. Completion of a task before due date may result in obsolete data, while completion after due date may lead to a serious failure. Just-in-time scheduling problems considered in the context of computer systems are discussed in the next section. They are referred to as real-time systems.

1.2 Computer systems

Similarity between scheduling problems and algorithms appearing in computer and manufacturing systems has been explored in theory and practice (see [33]). In the domain of computer systems the idea of just-in-time scheduling is most explicitly reflected in so-called real-time systems.

1.2.1 Real-time systems

A real-time system is any information processing system which has to respond to externally generated input stimuli within a finite and specified period. A correct performance of a task depends not only on the logical result but also the time the result is delivered. A failure to respond on time is as bad as giving an incorrect response.

Systems operating under time constraints occur in many application areas, such as embedded systems, vehicle control, industrial plant automation, robotics, multimedia audio and video stream conditioning, monitoring, and stock exchange orders. A good example of a real-time system is a keypad and monitor of a PC. A user must get visual feedback of each key stroke within a reasonable period. If the user cannot see that the key stroke has been accepted within this period the software product will at best be awkward to use. If, for example, the longest acceptable period was 100 ms, then any response between 0 and 100 ms would be acceptable. Another example may be the anti-lock brakes on a car. The real-time constraint in this system is the short time in which the brakes must be released to prevent the wheel from locking.

Real-time systems form a distinctive, very important part of contemporary computer science. Although real-time computing is ubiquitous, it is almost unnoticed. Many devices used every day are equipped with so called embedded systems. An embedded system is a special-purpose system in which the computer is completely encapsulated by the device it controls. Unlike a general-purpose computer, such as a personal computer, an embedded system performs one or a few pre-defined tasks, usually with very specific requirements. Since the system is dedicated to specific tasks, design engineers can optimize it, reducing the size and cost of the product. The user may only observe the features provided by the device like cellular phone, MP3 player, washing machine or refrigerator and is not aware of the sophisticated computer control system embedded inside the device. Hard real-time systems applied in embedded systems, typically interact at a low level with physical hardware. Other application of embedded systems include a car engine control system, medical systems such as heart pacemakers, industrial process controllers etc.

According to Butazzo [38], a typical real-time system (RTS):

- is reactive and interacts repeatedly with the environment;
- is usually a component of some larger application and is thus an embedded system which communicates via hardware interfaces;
- is trusted with important functions and thus must be reliable and safe;
- must perform multiple actions simultaneously, and be capable of rapidly switching focus in response to events in the environment, and therefore involve a high degree of concurrency;
- must compete for shared resources;
- its actions may be triggered externally by events in the environment or internally by the passage of time;
- should be stable when overloaded by unexpected inputs from the environment; completing important activities in preference to less important ones;
- is usually complex;
- is often long-lived (e.g. for decades in avionics applications) and thus must be maintainable and extensible;
- must be predictable, i.e. its timing behaviour is always within an acceptable range.

A ground station that processes data from a number of satellites may be an example of a real-time system. A different repetition interval is associated with each satellite. The satellite sends a message for a time period equal to its repetition interval regardless of whether or not the ground station receives the message. The station can receive a message from one satellite at a time. It should not miss any messages coming from all the satellites. The ground station scheduling problem is to assign each satellite at least one time slot for communication in each of its repetition intervals. This example illustrates two major features of real-time systems. The first one is that a task has to be completed within a time window; otherwise, a system failure occurs. The second one is that requests in real-time environments are often of recurring nature. Tasks that make requests at regular periodic intervals are called *periodic tasks*. In many real-time systems the period is a design parameter. Notice that it does not matter how long the period is between consecutive requests - a real-time system is not necessarily a high performance system. The most important feature of a real-time system is that missing the deadline means the system failure. The deadline is the latest time at which the task must complete. In our example it is the end of the repetition period for a given satellite. For periodic tasks, the deadline is usually the beginning of the next cycle.

Aperiodic tasks are also considered in real-time systems. Such tasks respond to randomly arriving events. For example, consider the anti-lock braking system in a car. When the driver presses the brake pedal, the anti-lock braking software actuates the brakes. Fast response is required. The main objective, however, is not to get the fastest response, but to specify the worst-case response time that still guarantees safe braking.

In order to define various types of real-time systems we can assign each task a profit function $z(C)$ that depends on the completion time C of the task. This function is defined on the interval $[r, \infty]$, where r is the earliest time at which processing of the task may start. Usually, the profit is constant in the interval $[r, dd]$, where dd is the deadline. Depending on the form of the profit function, three types of real-time systems are defined. The first one occurs when the profit has value $z(C)$, such that

$$z(C) = \begin{cases} a & \text{if } r < C < dd \\ -\infty & \text{if } C \geq dd. \end{cases} \tag{1.1}$$

It is called *hard real-time system*. The second type, called *firm real-time system*, differs in that the function $z(C)$ equals to

$$z(C) = \begin{cases} a & \text{if } r < C < dd \\ 0 & \text{if } C \geq dd. \end{cases} \tag{1.2}$$

Finally, in the third type, *soft real-time system*, a due date d is defined and it may be missed; however, the profit decreases if $d < C < dd$. Missing the deadline dd results in zero profit. The profit function is:

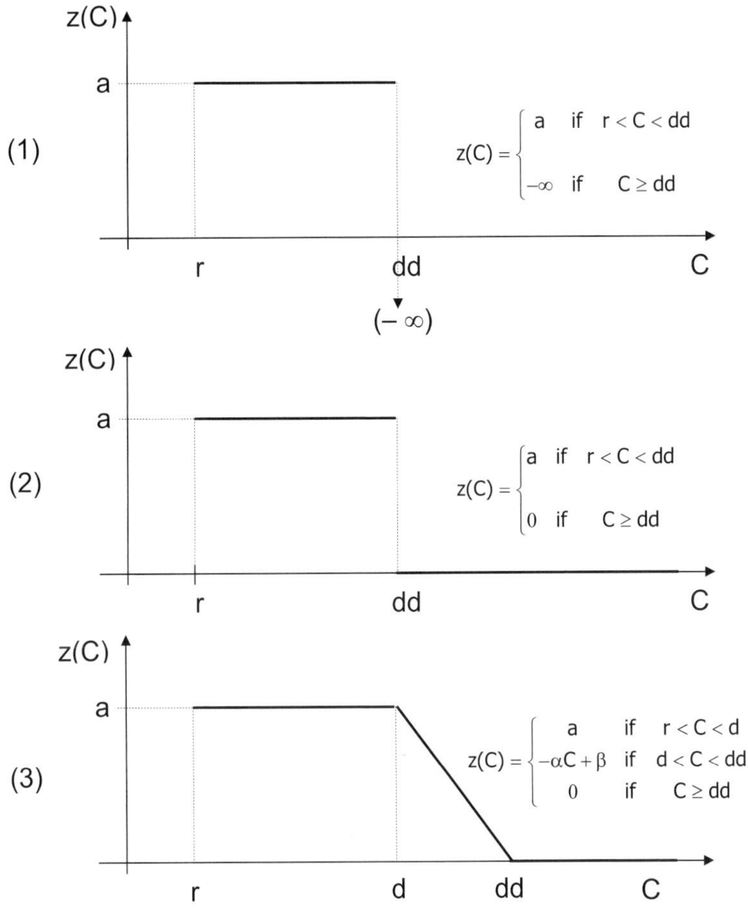

Fig. 1.5. Profit functions of real-time systems. Function $z(C)$ is a continuous function, $\alpha \geq 0$ and β are real numbers.

$$z(C) = \begin{cases} a & \text{if } r < C < d \\ -\alpha C + \beta & \text{if } d < C < dd \\ 0 & \text{if } C \geq dd. \end{cases} \tag{1.3}$$

Examples of the three types of the profit function are presented in Figure 1.5.

1.2.2 Hard real-time systems

Hard real-time systems are often considered mission-critical. This means that missing the deadline by one or more tasks constitutes complete failure of the system. Hard real-time constraints are required when the system failure causes a great loss in some manner, such as physically damaging the surroundings or threatening human lives. Systems that always have hard real-time constraints (due to the potentially severe outcome of missing a deadline) include nuclear power plants and car airbags.

For example, a temperature control system in a nuclear plant should monitor the temperature of the reactors and make appropriate decisions according to the value of the temperature. An important issue is to decide how often the control system should check the temperature. This frequency is called control rate and is defined by the control system designer. Appropriate control rate should be high enough to meet specifications, but not so high as to add unnecessary cost to the system. The control system must guarantee that an appropriate action is taken within a very well-defined time interval after the temperature value is exceeded. If a hard RTS is not able to meet these requirements it does not operate correctly, and the consequences of its malfunctioning may be so catastrophic that they are by no means tolerable.

The distinction between hard real-time and firm real-time systems is not always clear and usually not important from the point of view of the scheduling approach. Although it may be possible to assess whether the system is mission-critical or not, the result does not influence the scheduling problem. In both cases missing a deadline is forbidden and a schedule violating the deadlines is considered infeasible. Thus, further on we will consider hard real-time systems as described by either profit function of type (1.1) or (1.2). A class of algorithms solving one of the most common HRT scheduling problems, the Liu-Layland problem, will be examined in Chapter 5.

1.2.3 Soft real-time systems

In general, tasks in soft real-time systems are expected to meet the due dates. However, occasionally missing a due date is possible. It decreases the quality of service, but the system requirements as a whole continue to be met.

The examples of soft real-time systems include such applications as desktop audio and video, virtual reality, and Internet telephony.

As an example, let us consider a multimedia application that should provide the user with service of guaranteed quality. It follows from the technical specification that a frame of a movie must be delivered every 33ms. In order to assure desired quality, the corresponding speech should be presented not more than 80 ms later than the video frame is displayed. In practice, each new video stream to be transmitted by a network is tested for acceptance. If the network cannot satisfy the timing constraints of the stream without violating the constraints of existing streams, the new stream is rejected, and its admission is requested again at some later time. A user may accept a few glitches occurring rarely and for short lengths of time, especially taking into account that eliminating the glitches completely would significantly increase the cost of the device. Moreover, users of such systems rarely demand any proof that the system indeed honors its guarantees. For these reasons, the timing constraints of multimedia systems are considered soft.

A common approach to soft real-time constraints is to guarantee the performance on a statistical basis (e.g., the average number of late/lost frames per minute is less than 2). Soft real-time systems are not subject of analysis in this book; we just mention them in order to point out the difference between soft and hard real-time systems.

2

Methodological background

We start this chapter with providing some basic concepts in deterministic scheduling theory (Section 2.1). In Section 2.2 we introduce the theory of apportionment, developed by Balinski and Young. This theory originated from the problem of finding a fair representation of states in the US House of Representatives. Some methods of apportionment may be also used to solve scheduling problems, especially schedule balancing problems. We present these applications in Chapter 5.

2.1 Deterministic scheduling theory

Although excellent literature on deterministic scheduling theory is available (see, for example, [73, 70, 33, 36, 172]) in Section 2.1.1 we recall some basic definitions and notation for completeness of the presentation. In Section 2.1.2, we discuss the optimality criteria involving the earliness and tardiness costs in more detail. We also present an outline of the computational complexity theory of combinatorial problems (Section 2.1.3).

2.1.1 Basic definitions

In order to formulate a scheduling problem, a finite set of resources $\mathcal{R}$, a finite set of tasks $\mathcal{J}$, the schedule type and *optimality criteria* have to be defined. In this book we mainly consider scheduling problems with a single optimality criterion (objective function).

Generally speaking, scheduling problems consist in assigning resources to tasks in time in such a way that all constraints are satisfied and the considered optimality criterion is optimized. The two main classes of scheduling problems are *project scheduling* and *machine scheduling* problems. Project scheduling problems may be considered more general, since no restrictions on the resource types and requirements are imposed, while in machine scheduling machines are considered as a distinguished resource. Each task requires for its processing at least one machine. In the classical formulation of a machine scheduling problem only one task can be performed by a machine at a time and a task may be processed by one machine at a time only. Although some more general models are considered in the literature, in this book we only address the classical formulation.

The simplest system considered in the scheduling theory consists of a single machine. If a system consists of more than one machine, a machine either is able to perform all the tasks from a given set or is dedicated to performing a subset of tasks only. Thus, two main types of machine systems are defined: *parallel machines* and *dedicated machines.* Three types of parallel machine systems are distinguished depending on the task processing time. If the task processing time is independent of the machine, the machines are called *identical.* If the task processing time depends on the machine but the same proportion of processing times is preserved on all machines, the machines are called *uniform.* In the most general case the time of processing a task by a machine is arbitrary and the machines are called *unrelated.*

If a machine may perform a subset of tasks only, it is called a *dedicated machine.* Three types of systems with dedicated machines are defined: *flow shop*, *open shop* and *job shop.* In such systems each task is defined as a set of operations. Each operation may be performed by only one machine and the number of operations of a task is assumed to be equal to the number of machines m. In an open shop the order in which the operations are performed is arbitrary. In flow shops and job shops the order in which operations are performed is fixed for each task. In a flow shop the j-th operation of each tasks is performed on machine j, $j = 1, \ldots, m$. In a job shop the assignment of machines to operations may be arbitrary, although it is fixed for each task.

In many situations additional resources are taken into account. The total available number of units of each resource is limited. Discrete or continuous resources may be considered. Discrete resources that can be allotted to a task in amounts being numbers from a given finite set only, and continuous resources my be allotted in amounts from a given interval.

The set of tasks is described by:

- partial order, denoted by $\prec$, and defined over a set of tasks, which represents the precedence constraints between tasks. $i \prec j$ means that task i has to be completed before the processing of task j starts. Precedence constraints are usually represented by an acyclic directed graph;

- matrix of nonnegative processing times $[p_{ij}]$, $i = 1, \ldots, n$, $j = 1, \ldots, m$, where p_{ij} denotes the processing time of task i on machine j;
- matrix of resource requirements $[R_k(i)]$, $i = 1, \ldots, n$, $k = 1, \ldots, s$, where $R_k(i)$ denotes the amount of the additional resource k required by task i;
- vector of ready times $[r_i]$, $i = 1, \ldots, n$, where r_i denotes the earliest moment when the processing of task i may start;
- vector of due dates $[d_i]$, $i = 1, \ldots, n$, where d_i denotes the latest moment when the processing of task i should be completed. If a task is not completed by its due date, additional cost is incurred;
- vector of deadlines $[dd_i]$, $i = 1, \ldots, n$, where dd_i denotes the latest moment when the processing of task i must be completed. If a task is not completed by its deadline, the schedule is infeasible. If a deadline is not defined, it is assumed to be infinitely large;
- vector of weights $[w_i]$, $i = 1, \ldots, n$, where the weight of task i is interpreted as the relative importance of the task from the point of view of the considered optimality criterion.

Two types of schedules are considered: *nonpreemptive* and *preemptive* ones. In a nonpreemptive schedule the processing of a task must not be interrupted before completion. In a preemptive schedule the processing of a task may be stopped and resumed later without causing any additional cost. Preemption is feasible in computer systems, but usually not in manufacturing systems.

A *schedule* is an assignment of resources from set $\mathcal{R}$ to tasks from set $\mathcal{J}$ such that the following conditions are satisfied:

- at every moment each machine is assigned to at most one task and each task is processed by at most one machine;
- task i is processed in the time interval $[r_i, dd_i]$;
- all tasks are completed;
- if the schedule is nonpreemptive then no task is interrupted; if the schedule is preemptive, the number of interruptions is finite;
- precedence relations and resource constraints are satisfied.

Schedules are usually presented using the so-called *Gantt charts*. An example of a Gantt chart representing a schedule is given in Figure 2.1.

The quality of a schedule may be evaluated by various optimality criteria. Let us denote by C_i the completion time of task i, $i = 1, \ldots, n$

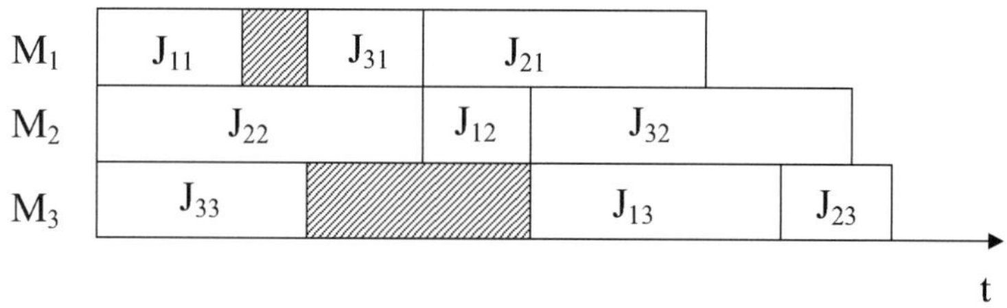

Fig. 2.1. A schedule of a set of tasks in an open shop. J_{ij} denotes operation j of task i.

and by $f_i(C_i)$ the cost of completing task i at time C_i. Basically, there are two types of optimality criteria: *bottleneck (maximum cost) objectives*, defined by formula (2.1), and *sum (total cost) objectives*, defined by formula (2.2).

$$\text{minimize} \max_{1 \leq i \leq n} \{f_i(C_i)\} \tag{2.1}$$

$$\text{minimize} \sum_{i=1}^{n} f_i(C_i) \tag{2.2}$$

The most common optimality criteria are the *makespan*, defined by formula (2.3) and the *mean flow time*, defined by formula (2.4). If $f_i(C_i) = w_i C_i$ in formulas (2.1) and (2.2) then the objectives are called weighted makespan and weighted mean flow time, respectively.

$$C_{max} = \max_{1 \leq i \leq n} \{C_i\} \tag{2.3}$$

$$F = \sum_{i=1}^{n} C_i \tag{2.4}$$

In the applications of scheduling in just-in-time and real-time systems one of the most important questions is the distance from the task due date d_i to task completion time C_i. For each task i we define:

- lateness

$$L_i = C_i - d_i, \tag{2.5}$$

- earliness

$$e_i = \max\{0, d_i - C_i\}, \tag{2.6}$$

- tardiness

$$t_i = \max\{0, C_i - d_i\}, \tag{2.7}$$

- unit penalty

$$U_i = \begin{cases} 0 \text{ if } C_i \le d_i \\ 1 \text{ otherwise.} \end{cases} \tag{2.8}$$

In the group of due date oriented optimality criteria, the most important bottleneck criterion is the maximum lateness

$$L_{max} = \max_{1 \le i \le n} L_i. \tag{2.9}$$

This criterion is often used in hard real-time systems. Its desired value is less than or equal to zero, meaning that all tasks are completed by their due dates. Of course, a weighted version of this criterion may be also considered and is defined as follows:

$$L_{max} = \max_{1 \le i \le n} w_i L_i. \tag{2.10}$$

Total tardiness and total weighted tardiness is defined, respectively, as follows:

$$T = \sum_{i=1}^{n} t_i, \tag{2.11}$$

$$T_w = \sum_{i=1}^{n} w_i t_i. \tag{2.12}$$

Another objective is minimization of the number of tardy tasks, or the weighted number of tardy tasks, calculated as follows:

$$U = \sum_{i=1}^{n} U_i, \tag{2.13}$$

$$U_w = \sum_{i=1}^{n} w_i U_i. \tag{2.14}$$

In just-in-time systems additional cost is associated with both tardy and early task completion. Thus the general optimality criterion involving total earliness and tardiness costs is formulated as follows:

$$f(S) = \sum_{i=1}^{n} (f_i^e(e_i) + f_i^t(t_i)). \tag{2.15}$$

where f_i^e, f_i^t are cost functions of earliness and tardiness, respectively. Usually they are non-decreasing real functions such that $f_i(0) = 0$.

Scheduling problems involving other optimality criteria as well as problems with multiple criteria are also considered in the literature. In this book, however, we are mainly concerned with due date oriented criteria. We discuss various objective functions involving the earliness and tardiness costs in Section 2.1.2.

A feasible schedule for which the value of the considered criterion is minimal is called an *optimal schedule.* Formally, a *scheduling problem* is defined as a set of parameters that characterize tasks and resources (including machines) together with the optimality criterion. Not all parameters need to be defined explicitly. For example, if nothing is stated about the precedence constraints, it is assumed that tasks are independent. An *instance* of a problem is obtained by specifying the values of all the problem parameters. Scheduling problems are usually formulated as optimization problems; however, sometimes decision problems are considered. An example of a scheduling problem in the decision version is a problem of finding a schedule in which no task is completed after its due date.

2.1.2 Earliness and tardiness cost functions

Two main approaches to modeling the just-in-time systems can be distinguished. The first one is applicable in the case of the make-to-stock production, where due dates are not explicitly defined, but the proportion of a particular product in total production should remain constant at all time. Mathematical models of scheduling problems and algorithms developed within this approach are presented in Chapter 5.

The second approach is more appropriate in the case of the make-to-order production. In this case a due date is assigned to each production order. An important class of earliness/tardiness scheduling problems includes problems with all the tasks sharing the common due date. Such situation takes place, for example, when several parts are required at the same time on the assembly line. Problems with a common due date are considered in Chapter 3 and problems with task dependent due dates in Chapter 4.

The objective of the earliness/tardiness scheduling is to minimize the earliness and tardiness costs. The earliness cost corresponds to the inventory costs and depends on the length of time the finished items are stored before shipment. The tardiness cost occurs if an order is completed after its due date, and corresponds to the penalties that have to be paid in such case. The main difficulty that arises within this approach is the formulation of the cost functions that best describe the relation between earliness/tardiness length and the corresponding inventory/late delivery cost. The most commonly used model assumes linear cost functions, but in some cases it is too simplistic. Tardiness costs are especially difficult to measure, because they encompass qualitative components e.g. customer dissatisfaction, and risk of losing a client. Nevertheless, this approach is very helpful in decreasing the overall production costs, even if the cost functions do not perfectly model the real costs. In this section we present the earliness and tardiness cost functions most commonly examined in the literature. Surveys of the scheduling problems with earliness/tardiness cost are presented in [217, 211, 60, 17, 145, 107] and [108]

Let us consider a set of n tasks. Task i is characterized by its processing time p_i and due date d_i, $i = 1, 2, \ldots, n$. The earliness e_i and tardiness t_i of task i, $i = 1, 2, \ldots, n$, are defined by equations (2.6) and (2.7), respectively. Two functions are associated with each task: the function of earliness cost f_i^e, and the function of tardiness cost f_i^t, $i = 1, 2, \ldots, n$.

Given a feasible schedule S, a vector of completion times $[C_1, C_2, \ldots, C_n]$ of tasks is known and the total earliness/tardiness cost is calculated according to formula (2.15).

It is easy to notice that in the case of scheduling problem with a common due date, processor idle time never occurs in an optimal schedule before the last task is completed. However, in the case of task dependent due dates it may be advantageous to hold production for some time in order to reduce the task earliness. A simple example of such situation is presented in Figure 2.2.

Example 2.1. Assume that we have to schedule two tasks, both with processing time of 6 units, and due dates $d_1 = 8$ and $d_2 = 17$, on a single machine.

If no idle time is allowed, the first task starts at time zero, and completes at $C_1 = 6$. Then the second task starts at time 6 and completes at time $C_2 = 12$. In the first schedule both tasks are completed before their due dates. The positive earliness cost is incurred, since $d_1 - C_1 = 2 > 0$ and $d_2 - C_2 = 5 > 0$. In the second schedule two idle time periods appear. As a result, both tasks are completed on time and the total earliness and tardiness cost is zero.

Obviously, inserting idle time results in the deterioration of machine utilization. The maximization of machine utilization usually conflicts with just-in-time optimization objectives such as low inventories and earliness/tardiness cost. Although the machine idle time cost is not

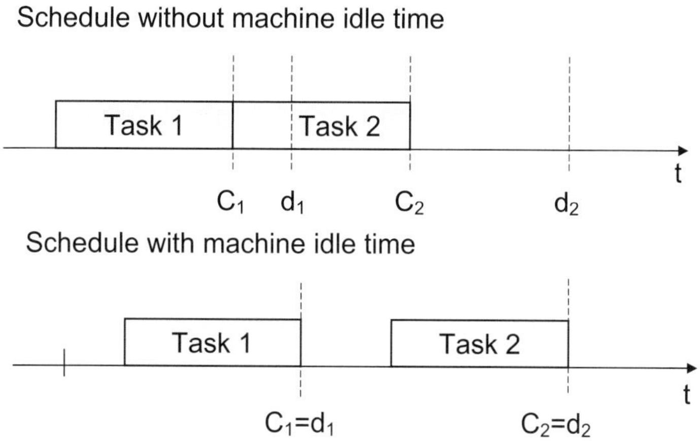

Fig. 2.2. Schedules with and without machine idle time.

negligible, it is consistent with the JIT philosophy to accept machine idle time in order to obtain lower earliness/tardiness cost. However, if the machine idle time cost is high, it may be justified to impose an additional constraint that no machine idle time may occur before completion of the last task scheduled. Thus, two main groups of problems are distinguished in the case of individual due dates of tasks: problems with idle time allowed and no idle time problems. The first case is more consistent with the idea of just-in-time scheduling, while the no idle time assumption may be justified in some other applications.

If an objective function is non-decreasing with respect to task completion times, it is called *regular*. Considering the earliness penalty usually results in an objective function that is not regular. The meaning of a regular and non-regular objective function is illustrated by the following example.

Example 2.2. Let $p_1 = 3, d_1 = 13, p_2 = 4, d_2 = 10$ and let the objective function be $f(S) = \sum_{i=1}^{2}(e_i + t_i)$. Consider two schedules presented in Figure 2.3. In the first schedule the completion times are $C_1 = 6, C_2 = 10$, and $f(S_1) = 7$. In the second schedule one of the values increases and we have $C_1 = 13, C_2 = 10$, while $f(S_2) = 0$. This example illustrates the fact that $f(S)$ is not a regular function.

Most of the scheduling problems with earliness and tardiness cost minimization are NP-hard, even if only one machine and a common due date is considered.

Let us comment on some special cases. We say that the earliness and tardiness costs are symmetric, if the cost of completing a task e_i

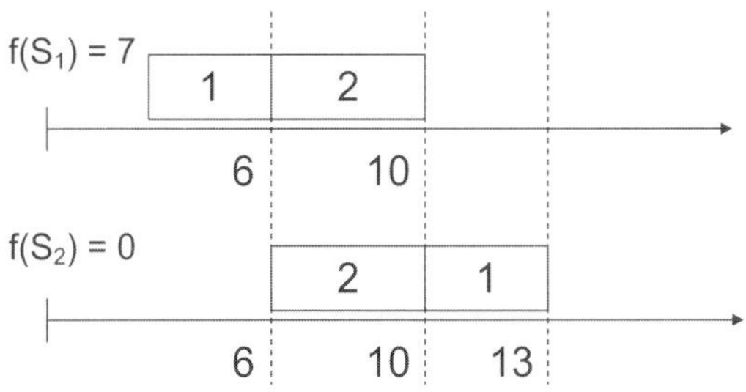

Fig. 2.3. Two schedules of tasks from Example 2.2.

time units before the due date is the same as completing it $t_i = e_i$ time units after the due date, i.e. $f_i^e = f_i^t$, $i = 1, 2, \ldots, n$. If the earliness and tardiness cost, respectively, is equal for all tasks, i.e. $f_i^e = f^e$ and $f_i^t = f^t$, $i = 1, 2, \ldots, n$, then we speak about task independent earliness and tardiness costs. It is easy to notice that if costs are symmetric and task independent, they are identical, i.e. $f_i^e = f_i^t = f$, $i = 1, 2, \ldots, n$.

The most intensively examined earliness/tardiness problems are the problems where f_i^t and f_i^e, $i = 1, 2, \ldots, n$, are linear functions. Linear functions represent situations where earliness or tardiness costs are proportional to the length of the earliness or tardiness, respectively. It is not always the case, since clients usually accept small tardiness, but expect large compensation if they must wait long for the product or service. Costs that grow according to this pattern of client behavior are better represented by a quadratic function. Interesting results are presented in the literature for linear and quadratic cost functions; however, only a few results are generalized to address arbitrary cost functions. In Table 2.1 we present the formulation of various earliness/tardiness optimality criteria. The coefficients α_i, β_i, $i = 1, 2, \ldots, n$, are arbitrary integers.

Scheduling problem with objective function given by formula (2.15) can be generalized by extending the objective function to include other measures of performance. For example, except for linear earliness/tardiness penalties it may be assumed that task i incurs a completion time penalty $\theta_i C_i$, where θ_i, $i = 1, \ldots, n$, is an arbitrary real coefficient. The objective function has the following form

$$\sum_{i=1}^{n} (\alpha_i e_i + \beta_i t_i + \theta_i C_i) \tag{2.16}$$

Table 2.1. Formulation of various objective functions with earliness and tardiness costs

Cost functions	Linear	Quadratic	Arbitrary
General	$\sum_{i=1}^{n}(\alpha_i e_i + \beta_i t_i)$	$\sum_{i=1}^{n}(\alpha_i e_i^2 + \beta_i t_i^2)$	$\sum_{i=1}^{n}(f_i^e(e_i) + f_i^t(t_i))$
Symmetric	$\sum_{i=1}^{n}\alpha_i(e_i + t_i)$	$\sum_{i=1}^{n}\alpha_i(e_i^2 + t_i^2)$	$\sum_{i=1}^{n}f_i(e_i + t_i)$
Task dependent	$\sum_{i=1}^{n}(\alpha e_i + \beta t_i)$	$\sum_{i=1}^{n}(\alpha e_i^2 + \beta t_i^2)$	$\sum_{i=1}^{n}(f^e(e_i) + f^t(t_i))$
Task independent	$\sum_{i=1}^{n}\alpha(e_i + t_i)$	$\sum_{i=1}^{n}\alpha(e_i^2 + t_i^2)$	$\sum_{i=1}^{n}f(e_i + t_i)$

Another objective function is analyzed by Sidney [218]. He considers a single machine problem where each task is supposed to be performed in a given time window $[r_i, d_i]$, $i = 1, \ldots, n$. Task earliness is defined as $e_i = max\{0, r_i - C_i + p_i\}$ and the objective is to minimize the function

$$f(S) = \max_i\{f^e(e_i), f^t(t_i)\} \tag{2.17}$$

where f^e and f^t are convex, continuous and non-decreasing functions.

In some formulations it is assumed that also due dates are decision variables. Treating due dates as decision variables is relevant in the situation of setting due dates internally, as targets to guide the progress of shop floor activities.

Comprehensive surveys of results concerning the due date assignment problem are presented in [107] and [108]. The basis for classification of the due date assignment problems is the way in which the due dates are determined. This classification is presented in Table 2.2.

In the simplest problem, called CON (for *constant flow allowance*) each task has the same due date d. If a due date of task j is defined as $d_j = p_j + q$, where q reflects the waiting time or slack, equal for all tasks, the problem is called SLK (for *slack*). The problem where the due dates are related to the total work content, i.e. $d_j = vp_i$, where v is a coefficient common for all tasks, is denoted by TWK (for *total work*).

Table 2.2. Classification of due date assignment problems

Acronym	Explanation	Due date
CON	Constant flow allowance	$d_i = d$
SLK	Slack	$d_i = p_i + q$
TWK	Total work	$d_i = vp_i$
PPW	Processing plus wait	$d_i = vp_i + q$
NOP	Number of operations	$d_i = vn_i$

q and v are coefficients
n_i is the number of operations of task i

Another problem combines the rules for SLK and TWK resulting in $d_j = vp_j + q$. It is called PPW (from *processing plus wait*). Finally, in the NOP problem (for *number of operations*), due dates are determined on the basis of the number of operations n_j of task j, i.e. $d_j = vn_j$.

Concluding, the examination of the earliness and tardiness costs simultaneously resulted in numerous interesting scheduling problems, extending the models and methods of the classical scheduling theory.

2.1.3 Scheduling algorithms and computational complexity

A *scheduling algorithm* is an algorithm which constructs a schedule for any instance of the given problem. An *optimization algorithm* is an algorithm that finds an optimal schedule for any problem instance. An algorithm that may fail to find an optimal schedule for some problem instance is called a *heuristic algorithm* or a *heuristic.*

We call an approximation algorithm guaranteed to find a solution at most (or at least, as appropriate) ε times the optimum a ε-*approximation algorithm.* The ratio ε is called the performance ratio or *relative performance guarantee* of the algorithm.

An *approximation scheme* is an important class of heuristic algorithms. It is a set of algorithms $\{A(\varepsilon)|\ \varepsilon > 0\}$, where each $A(\varepsilon)$ is a $(1+\varepsilon)$-approximation algorithm and the execution time is bounded by a polynomial in the length of the input. The execution time may depend on the choice of ε. More properly, approximation scheme is called a polynomial-time approximation scheme (PTAS).

The characteristics of a scheduling algorithm consists in determining its performance guarantee and computational complexity. For many algorithms, it is difficult to establish the performance guarantee theoretically, so it is estimated on the basis of computational experiments. Thus, usually, instead of the worst-case performance the result of the experiment validates the mean performance of an algorithm.

We are always interested in solving a problem in the shortest possible time. It is also the case when scheduling problems are considered. Moreover, in some applications (e.g. hard real-time systems) the solution time may be crucial for the system survival, since if a solution is found too late, the system will fail to perform its main function. Thus, the analysis of time complexity of scheduling algorithms is very important not only from the theoretical, but also from the practical point of view.

We refer the reader to [103] for a comprehensive study of the complexity analysis of combinatorial problems. For the purpose of this book we only define two main classes of problems: $\mathcal{P}$ and $\mathcal{NP}$.

The analysis of computational complexity requires a well defined model of computation. Most often, the Turing machine is used as a mathematical model of an algorithm. An algorithm may be viewed as a function which maps each instance of a problem (input) onto a solution (output). If the output is in the set $\{yes, no\}$, the problem is called a *decision problem*. We can define the *time complexity function* of an algorithm as a function that assigns each input length n the maximum number of steps $T(n)$ in which the algorithm finds a solution for any input of length n. Usually, the precise form of T is replaced by its asymptotic order where the notation $T(n) \in O(g(n))$ means that there exist a constant $c > 0$ and a nonnegative number n_0 such that $T(n) \leq cg(n)$ for all integers $n \geq n_0$.

The computational complexity of an algorithm depends on the encoding of the input. If the instance is represented using binary encoding, we say that an algorithm is *polynomial* if it solves the problem in polynomial time, i.e. if there exists a polynomial $p(n)$ such that $T(n) \in O(p(n))$, where n is the input length of an instance. An algorithm is called *pseudopolynomial* when $T(n) \in O(p(n))$ if n is the input length with respect to unary encoding. An algorithm is called *exponential time algorithm* if its complexity function cannot be bounded by a polynomial of n.

The class of decision problems for which polynomial algorithms exist is denoted by $\mathcal{P}$. If a positive answer (*yes*) to a decision problem can be verified in polynomial time, this problem belongs to the class $\mathcal{NP}$. Is is easy to notice that $\mathcal{P} \subseteq \mathcal{NP}$. It is an open question, however, if $\mathcal{P} = \mathcal{NP}$.

For two decision problems Π and Θ, we say that Π reduces to Θ (denoted $\Pi \propto \Theta$) if there exists a polynomial-time computable function that transforms any input x for Π to an input y for Θ such that the output for x is *yes* iff the output for y is *yes*. A decision problem Θ is called *NP-complete* if $\Theta \in \mathcal{N}P$ and if for any other decision problem $\Pi \in \mathcal{N}P$ we have $\Pi \propto \Theta$.

There are many scheduling problems for which polynomial algorithms are not known. In fact, it is an open question if such algorithms exist for these problems. It is important to notice that if any NP-complete problem could be solved in polynomial time, then all problems in $\mathcal{N}P$ could be solved in polynomial time. Thus, if for any problem a polynomial time algorithm is not known, an important result is to prove that this problem is NP-complete. The satisfiability problem from the Boolean logic is the first problem shown to be NP-complete (see [74]). Now, proving that a problem is NP-complete consists in finding a reduction of any known NP-complete problem to the examined one.

Let us denote by $\Pi|_q$ a subproblem of a decision problem Π, obtained by restricting Π to instances of length $q(n)$, where n is the length of the input in unary encoding and q is a polynomial. The decision problem Π is *NP-complete in the strong sense* (strongly NP-complete) if $\Pi \in \mathcal{N}P$ and if there exists a polynomial q defined for integers for which $\Pi|_q$ is NP-complete.

An optimization problem whose decision version is NP-complete is called *NP-hard*. Although it is very unlikely that polynomial time algorithms exist for NP-hard problems, branch-and-bound algorithm or other enumeration algorithms may be applied to solve small instances. Moreover, for ordinary NP-hard problems pseudopolynomial time algorithms may be constructed. Finally, the most common approach is do develop heuristic algorithms for NP-hard problems. The heuristics should be of polynomial complexity and should provide solutions close to optimal ones.

2.2 The Theory of Apportionment

The *problem of apportionment* appears in many practical situations. The most typical one is the assignment of seats in a house of representatives. Usually, each state or department represented in the parliament receives a number of seats that is proportional to its population but not fewer than a given minimum number of seats (this number is regulated by law and differs for different parliaments). This problem is most comprehensively examined by Balinski and Young in [20].

Bautista, Companys and Corominas [22] were the first to observe the relations between the Product Rate Variation problem and the apportionment problem. These relations allow to prove many interesting results concerning the properties of scheduling algorithms developed for the PRV problem [134]. Later, a transformation between the Liu-Layland and the PRV problem was shown [156]. A simple conclusion was that the Liu-Layland problem may be transformed to the apportionment problem. Based on this transformation further properties of the scheduling algorithms were examined.

Since the problem of apportionment is equivalent to two just-in-time scheduling problems, namely, the PRV problem and the Liu-Layland problem, in this section we present an introduction to the theory of apportionment.

2.2.1 Problem formulation

The apportionment problem is formulated as follows. Given n states, a vector of their populations $\boldsymbol{\pi} = [\pi_1, \ldots, \pi_n]$, the house size h, and a vector of minimal requirements $\boldsymbol{\varrho} = [\varrho_1, \ldots, \varrho_n]$, find an *apportionment*, i.e. a vector of nonnegative integers $\mathbf{a} = [a_1, \ldots, a_n]$, such that $a_i \geq \varrho_i$, $i = 1, \ldots, n$, and $\sum_{i=1}^{n} a_i = h$.

A *method of apportionment* $M(\boldsymbol{\pi}, h)$ is a multi-valued function M, consisting of a set of apportionments of h among n, for each n-vector $\boldsymbol{\pi} > 0$ and integer $h \geq 0$. A single-valued function $f(\boldsymbol{\pi}, h) = \mathbf{a} \in M(\boldsymbol{\pi}, h)$ that breaks every tie in some arbitrary fashion is a particular M – *solution*. A tie is a situation where a method gives more than one apportionment for the same problem. For example, if we have two states with populations $\pi_1 = \pi_2$, and the number of seats to be assigned is $h = 3$, a certain method may find two apportionments, $\mathbf{a} = [1, 2]$ and $\mathbf{a}' = [2, 1]$. Breaking a tie means choosing one of the solutions, $\mathbf{a}$ or $\mathbf{a}'$. Various types of ties may arise, depending on the method of apportionment used.

Below we recall the most important concepts and properties of the classical methods of apportionment.

A natural requirement is to find a fair method of apportionment, i.e. one that assigns each state a number of seats a_i proportional to its population $\boldsymbol{\pi}_i$. The ideal number of seats for state i, called *quota*, is defined as follows:

$$q_i = \frac{\pi_i h}{\sum_{k=1}^{n} \pi_k}. \tag{2.18}$$

Obviously, this ideal can rarely be met because the quota may not be integer. Thus, the first issue that has to be addressed is how to characterize a fair method of apportionment. In the attempt to answer this question, Balinski and Young [20] provide a systematic analysis of the methods used so far.

It is natural to expect that a fair method of apportionment should find the same apportionment $\mathbf{a}$ of seats in a parliament of size h for vectors $\boldsymbol{\pi}$ and $\lambda\boldsymbol{\pi}$, where λ is any positive rational number. A method which has this property is called *homogeneous*. Moreover, any method of apportionment should be *symmetric*, i.e. it should find the same apportionment of seats in a house of size h for any vector obtained by permutation of elements in vector $\boldsymbol{\pi}$. Another property, called *proportionality*, means that whenever all quotas q_i, $i = 1, \ldots, n$, are integers, the method finds only the solution $a_i = q_i$, $i = 1, \ldots, n$. The definition of a method of apportionment may be extended to the case where population sizes are real numbers. A method of apportionment is *complete* if the following property holds:

$$\left(\boldsymbol{\pi}^n \to \boldsymbol{\pi} > \mathbf{0} \wedge \forall n \;\; \mathbf{a} \in M(\boldsymbol{\pi}^n, h)\right) \Rightarrow \left(\mathbf{a} \in M(\boldsymbol{\pi}, h)\right).$$

All methods considered later in this chapter are homogeneous, symmetric, proportional, and complete.

A fair method of apportionment should ensure that if the house size increases, no state loses a seat in the house. A method with this property is called *house monotone*. A method M is house monotone if for any population $\boldsymbol{\pi}$ and house h, $\mathbf{a} \in M(\boldsymbol{\pi}, h)$ implies $\mathbf{a}' \in M(\boldsymbol{\pi}, h + 1)$ for some $\mathbf{a} \geq \mathbf{a}'$. If a method of apportionment does not possess this property, the so-called Alabama paradox may occur. The Alabama paradox was the first of the apportionment paradoxes to be discovered. In 1880, C. W. Seaton, chief clerk of the U. S. Census Office, used the *Hamilton method* to calculate the apportionments for two sizes of parliament: 299 and 300 seats. He showed that Alabama would get 8 seats in the first one, and only 7 seats in the second, bigger parliament. Clearly, the Hamilton method is not house monotone.

The Hamilton method works as follows.

Algorithm 2.3 (Hamilton method [20]).

1. Assign each state a number of seats equal to its quota rounded down to an integer.

2. For each state calculate the difference between its quota and its quota rounded down to an integer.
3. Assign the remaining seats to states with the biggest differences calculated in the previous step.

A fair method of apportionment should also meet the requirement that if the population of a state increases, then its representation in the parliament should not decrease. This property is called *population monotonicity*. Formally, a method is population monotone if for any two vectors of populations $\boldsymbol{\pi}, \boldsymbol{\pi}' > 0$ and vectors of apportionments $\mathbf{a} \in M(\boldsymbol{\pi}, h)$, $\mathbf{a}' \in M(\boldsymbol{\pi}', h')$ the following implication holds:

$$\frac{\pi'_{i'}}{\pi'_{j'}} \geq \frac{\pi_i}{\pi_j} \Rightarrow \begin{cases} (a'_{i'} \geq a_i \vee a'_{j'} \leq a_j) \\ \text{or} \\ \dfrac{\pi'_{i'}}{\pi'_{j'}} = \dfrac{\pi_i}{\pi_j} \quad \text{and} \\ a'_{i'}, a'_{j'} \text{ can be substituted for } a_i, a_j \text{ in } \mathbf{a}. \end{cases} \tag{2.19}$$

Balinski and Young [20] show that any population monotone method is house monotone but not vice versa.

2.2.2 Divisor methods

An important class of apportionment methods, the so-called *divisor methods*, is based on a specific rounding procedure. For any positive real number z, a *d-rounding* of $[z]_d$ is an integer a such that $d(a-1) \leq d \leq d(a)$, which is unique unless $z = d(a)$, in which case the d-rounding takes either of the values a or $a+1$. Any monotone increasing $d(a)$ defined for all integers $a \geq 0$ and satisfying $a \leq d(a) \leq a+1$ is called a *divisor criterion*. The *divisor method* based on d is defined as follows:

$$M(\boldsymbol{\pi}, h) = \{\mathbf{a} : a_i = [\pi_i/x]_d \text{ and } \sum_i a_i = h \text{ for some } x\}. \tag{2.20}$$

Obviously, there is an infinite number of different divisor methods. Table 2.3 shows the divisor criteria $d(a)$ of the traditional divisor methods. Any divisor method can be defined as an iterative procedure. Below we present the divisor method in this form, assuming that population vector $\boldsymbol{\pi}$ and house size h are given.

Table 2.3. The traditional divisor methods

Divisor method	Divisor criterion $d(a)$
Adams	a
Dean	$\frac{a(a+1)}{a+1/2}$
Hill	$\sqrt{a(a+1)}$
Webster	$a+\frac{1}{2}$
Jefferson	$a+1$

Algorithm 2.4 (Divisor method [20]).

1. Set $a_i = 0$ for $i = 1, \ldots, n$.
2. Set $k = 1$.
3. Calculate $\pi_i/d(a_i)$ for $i = 1, \ldots, n$.
4. Find j such that $\pi_j/d(a_j) = \max_i \{\pi_i/d(a_i)\}$.
5. Set $a_j = a_j + 1$.
6. Set $k = k + 1$.
7. If $k \leq h$ then go to step 3, else stop.

One of the most interesting divisor methods is the Webster method. It is the only divisor method which is both proportional and unbiased, i.e. it favors neither small nor large states. We illustrate the Webster method with an example.

Example 2.5. Assume that 5 seats should be apportioned between three states with populations 7270, 4230 and 2220.

Table 2.4. Solution of the problem from Example 2.5

State	A	B	C
k	$\pi_i/(a_i+0.5)$		
1	**14540**	8460	4440
2	4847	**8460**	4440
3	**4847**	2820	4440
4	2908	2820	**4440**
5	**2908**	2820	1480
a	3	1	1

In step 3 of Algorithm 2.4 values $\pi_i/(a_i(h)+0.5)$ are calculated for each house size $k \leq h$ and each state A, B, and C. In each iteration a seat is allotted to the state with the maximum value of the divisor criterion. The results obtained for Example 2.5 are presented in Table 2.4. The values in boldface indicate which state receives the seat in each iteration. The ultimate solution is $\mathbf{a} = [3, 1, 1]$.

Balinski and Young prove in [20] that an apportionment method is a divisor method if and only if it is population monotone. Thus, any divisor method is also house monotone.

Divisor methods with $d(a) = a + \delta$, where $0 \leq \delta \leq 1$, are called *parametric methods*. Thus, Adams, Webster and Jefferson methods are parametric methods. Parametric methods are cyclic, i.e. if a method finds an apportionment for population $\boldsymbol{\pi}$ and house size h, then the same method applied to population $\lambda\boldsymbol{\pi}$ and parliament λh assigns first h seats to states in the same order as in the original problem, and then it repeats this order λ times.

An apportionment method is *uniform* if for every $l, 2 \leq l \leq n$, $(a_1, \ldots a_n) \in M((\pi_1, \ldots, \pi_n), h)$ implies $(a_1, \ldots a_l) \in M((\pi_1, \ldots, \pi_l), \sum_1^l a_i)$, and, if also $(b_1, \ldots b_l) \in M((\pi_1, \ldots, \pi_l), \sum_1^l a_i)$ then $(b_1, \ldots b_n) \in M((\pi_1, \ldots, \pi_n), h)$. Intuitively speaking, uniformity means that every restriction of a fair apportionment is fair. Moreover, if a restriction yields a different apportionment of the same number of seats, then there is a corresponding tie in the original, unrestricted problem.

Uniform methods are also called *rank-index methods* because any uniform method can be obtained using the so-called *rank-index*. A rank-index $\varphi(\pi, a)$ is any real-valued function of rational π and integer $a \geq 0$ that is decreasing in a, i.e. $\varphi(\pi, a-1) > \varphi(\pi, a)$. The rank-index method based on the rank-index φ is defined as follows.

Algorithm 2.6 (Rank-index method[20]).

1. Set $a_i = 0$ for $i = 1, \ldots, n$.
2. Set $k = 1$.
3. Calculate $\varphi(\pi_i, a_i)$ for $i = 1, \ldots, n$.
4. Find j such that $\varphi(\pi_j, a_j) = \max_i \{\varphi(\pi_i, a_i)\}$.
5. Set $a_j = a_j + 1$.
6. Set $k = k + 1$.
7. If $k \leq h$ then go to step 3, else stop.

Divisor methods are a subclass of the rank-index methods in which the rank-index has the form $\varphi(\pi, a) = \pi/d(a)$, so every divisor method is uniform. Since each uniform method is rank-index, it is also house monotone.

2.2.3 Staying within the quota

In addition to the properties discussed in the previous section, it is reasonable to require that any state is allotted the number of seats that is equal to at least its quota rounded down and at most its quota rounded up. This property is called *staying within the quota*. More formally, an allotment of a_i seats to state i stays above the *lower quota* for population π and house size h if and only if

$$a_i \geq \left\lfloor \frac{\pi_i h}{\sum_{k=1}^{n} \pi_k} \right\rfloor \tag{2.21}$$

and it stays below the *upper quota* if and only if

$$a_i \leq \left\lceil \frac{\pi_i h}{\sum_{k=1}^{n} \pi_k} \right\rceil. \tag{2.22}$$

An apportionment vector $\mathbf{a}$ *stays within the quota* if it simultaneously stays above the lower and below the upper quota for all states. Intuitively, an apportionment stays within the quota if it deviates from its quota by less than one seat. A method of apportionment stays within the quota, if every apportionment obtained by this method stays within the quota.

Some house monotone methods stay within the quota. The first such method is the *Quota method* proposed by Balinski and Young in [19]. Still in [224] proposes a more general method with the same property. The main idea of both methods is to assign the consecutive seat to a state only in the case if this assignment does not violate the quota. The eligibility tests performed in the Still method examine the possibility to violate the quota even a few steps ahead.

The eligible set $E(h)$ for house size h consists of all the states which may be allotted the available seat without violating the quota, either for h or any bigger house size. For any house size $h > 0$, the eligible set $E(h)$ consists of all states i that pass the eligibility test described below.

Denote by a_{ih} and by q_{ih} the number of seats and the quota of state i in a house of size h, respectively. The goal of the test is to create a set of states eligible to receive the consecutive seat if the size of the parliament grows by one seat. Let h_i be the house size at which state i first becomes entitled to obtain the next seat, i.e. h_i is the smallest house size greater than or equal to h at which the lower quota of state i is greater than or equal to $a_{ih} + 1$, or equivalently

$$h_i = \left\lceil \frac{a_{ih}+1}{\pi_i} \sum_{i=1}^{n} \pi_i \right\rceil . \tag{2.23}$$

Algorithm 2.7 (Eligibility tests for state i and house size h [224]).

1. *The upper quota test.* If the number of seats $a_{ih} \geq \lceil q_{i(h+1)} \rceil$ then the state is not eligible to receive the next seat. Stop.
2. *The lower quota test.*
   a) For each house size $g \in (h, h_i)$ set $s_i(g, i) = a_{ih}+1$ and $s_j(g, i) = \max\{a_{ih}, \lfloor q_{jg} \rfloor\}$, for $j \neq i$.
   b) If for some house size $g \in (h, h_i)$, $\sum_{j=1}^{n} s_j(g, i) > g$, then state i is not eligible to receive the next seat. Stop.

The eligibility test is performed for all states and the states that are not rejected by the upper or lower eligibility test belong to set $E(h+1)$. Still proved in [224] that the eligible set is never empty. The Still method assigns the consecutive seat to a state with the highest priority from the eligible set.

Algorithm 2.8 (Still method [224]).

1. For the house of size $h = 0$ assign each state 0 seats.
2. For each k, $k = 1, ..., h$, construct eligible set $E(k)$ and assign the k-th seat to the state with the highest priority of all states from the eligible set $E(k)$.

Usually, performing the eligibility test requires more computation than calculating the priorities, so it may be more efficient to select the state with the highest priority first and then perform the eligibility test for this state. If the state does not pass the test (which is a rare situation), it is removed from the set of states and the state with the highest priority from the reduced set is tested. As we have stated above, at least one state passes the eligibility test.

The Still method represents the algorithms called *quota methods.* Any house monotone method staying within the quota belongs to this class. Algorithms of this class differ in the way they assign priorities to states from the eligible set. Any divisor method may be used to define priorities of states in the eligible set $E(h)$. A quota method that assigns priorities to tasks using a divisor method is called a *quota-divisor method* and works as follows.

Algorithm 2.9 (Quota-divisor method [20]).

1. Set $a_i = 0$ for $i = 1, \ldots, n$.
2. Set $k = 1$.

3. Calculate $\pi_i/d(a_i)$ for $i = 1, \ldots, n$.
4. Find j such that $\pi_j/d(a_j) = \max_i \{\pi_i/d(a_i)\}$ and $j \in E(k)$.
5. Set $a_j = a_j + 1$.
6. Set $k = k + 1$.
7. If $k \leq h$ then go to step 3, else stop.

The quota-divisor methods stay within the quota, while the divisor methods do not.

We illustrate the Still method with an example. We use the Adams method for ranking the states in the eligible set.

Example 2.10. Let us consider the problem of apportionment of 6 seats among 4 states, A, B, C and D, with populations 1500, 3000, 6000, and 7000, respectively. According to the Adams method, the state that should receive the consecutive seat is state D, because we choose the state with the biggest value $\pi_i/1$. We have to check if state D is eligible. State D passes the upper quota test, since initially $a_{40} = 0$, while the corresponding upper quota equals 1. Let us now perform the lower quota test for state D ($i = 4$). The house size h_4 in which state D is eligible to receive the second seat, calculated according to formula (2.23), equals 3. We present the values $s_j(g, 1)$ for $j = 1, 2, 3, 4$, and $g \in \{1, 2, 3\}$ in Table 2.5.

Table 2.5. Lower quota test for $k = 1$ and $i = 4$

	$\lfloor q_{jg} \rfloor$		
j	1	2	3
1	0.09	0.17	0.26
2	0.17	0.34	0.51
3	0.34	0.69	1.03
	$s_j(g,1)$		
j	1	2	3
1	0	0	0
2	0	0	0
3	0	0	1
4	1	1	1
$\sum_{j=1}^{4} s_j(g,1)$	1	1	2

State D passes the lower quota test and it is eligible to receive the first seat, so after the first iteration $a_{11} = 0, a_{21} = 0, a_{31} = 0$, and $a_{41} = 1$. In the second iteration state C has the highest priority, so we perform its eligibility test. Since $a_{31} = 0$ and $q_{31} = 0.34$, the state passes the upper quota test. For the lower quota test we find the house size in which state C is eligible to receive the next seat, $h_3 = 3$. The test data for $g \in \{2, 3\}$ are presented in Table 2.6.

Table 2.6. Lower quota test for $k = 2$ and $i = 3$

	$\lfloor q_{jg} \rfloor$	
j	2	3
1	0.17	0.26
2	0.34	0.51
4	0.80	1.20
	$s_j(g,1)$	
j	2	3
1	0	0
2	0	0
3	1	1
4	1	1
$\sum_{j=1}^{4} s_j(g,2)$	2	2

The second seat is assigned to state C, so $a_{12} = 0, a_{22} = 0, a_{32} = 1$, and $a_{42} = 1$. According to the Adams method, the third seat should be assigned to state D. State D is eligible to get this seat, so $a_{13} = 0, a_{23} = 0, a_{33} = 1$, and $a_{43} = 2$. We have a tie while assigning the fourth seat, since values π_i/a_{i4} are equal 1500, 3000, 3000, and 2333.3, for $i = 1, 2, 3$, and 4, respectively. We choose arbitrarily, for example state B. Proceeding this way we finally obtain: $a_{16} = 0, a_{26} = 1, a_{36} = 2$, and $a_{46} = 3$.

Lastly, let us show that the *Quota method* proposed by Balinski and Young in [19] may be considered a special case of the method proposed by Still. Namely, observe that the eligibility test of the *Quota method* consists only of the upper quota test of the Still method, and the Jefferson divisor criterion $d(a) = a + 1$ is used to rank the eligible states.

2.2.4 Impossibility Theorem

In Sections 2.2.2 and 2.2.3 we presented two classes of methods of apportionment, i.e. the divisor methods and methods that stay within the

Table 2.7. Classification of the methods of apportionment

	House monotone methods		Methods that are not house monotone
	Population monotone methods	Methods that are not population monotone	
Methods staying witin the quota	Empty set (see the Impossibility Theorem)	Still method	Hamilton method
Methods not staying witin the quota	Uniform = Rank-index methods Divisor methods (Jefferson, Adams, Webster, Dean, Hill method)		

quota. In this section we show that these classes are disjoint. To this end we use the famous *Impossibility Theorem*, proved by Balinski and Young [20], and saying that no method can be population monotone and stay within the quota at the same time.

Theorem 2.11 (Impossibility Theorem [20]). *No method of apportionment exists for $n \geq 4$ and $h \geq n+3$ that is population monotone and stays within the quota.*

We mentioned in Section 2.2.2 that each divisor method is population monotone. Thus, an immediate conclusion from the Impossibility Theorem is that no divisor method stays within the quota. Moreover, methods that stay within the quota may be house monotone; however, they are not population monotone. Thus, quota-divisor methods are not population monotone, although the divisor methods are. Moreover, Balinski and Young prove in [20] that every divisor method stays within the quota for all 2-state problems and that Webster method is the only divisor method that stays within the quota for all 3-state problems.

Divisor methods have another important property, stated in Theorem 2.12.

Theorem 2.12 (Balinski and Young [20]). *No apportionment obtained by a divisor method simultaneously violates the upper quota of one state and violates lower quota of another.*

Although no divisor method stays within the quota for any population vector, some of them, for example the Webster method, violate the quota very rarely. Moreover, any apportionment obtained by the Webster method has the property that no state can be brought closer to its quota without moving another state further from its quota.

The classification of the methods of apportionment based on the Impossibility Theorem is presented in Table 2.7.

3

Common due date

In this chapter we consider the scheduling problems with minimization of the earliness and tardiness costs, where all tasks should be completed at the same due date. Problems with minimization of the earliness and tardiness costs occur in just-in-time systems as we discuss in Section 1.1.4. Considering a common due date is relevant in the situation where several items constitute a single order, or they are delivered to an assembly line in which the components should all be ready at the same time to avoid storing inventories or delays of deliveries. Since this requirement cannot be fulfilled the scheduling objective is minimization of the earliness and tardiness costs.

We consider the problem of scheduling n independent tasks with processing times p_i, $i = 1, \ldots, n$, and a common due date d. The earliness and tardiness of task i are calculated as follows: $e_i = \max\{d - C_i, 0\}$ and $t_i = \max\{C_i - d, 0\}$, $i = 1, \ldots, n$, respectively.

If not stated otherwise, we will assume that non-preemptive tasks, with zero ready times and arbitrary processing times are considered.

The majority of results on the earliness/tardiness scheduling are concerned with single machine systems. Some polynomially solvable cases are formulated for the parallel uniform and unrelated machines. The problems with precedence constraints are rarely considered in the context of machine earliness/tardiness scheduling problems. However, the project scheduling problems with earliness/tardiness criteria are also examined, e.g. in [241].

Usually, the due date is a problem parameter since it results from the customer order. However, the delivery date may be often negotiated with the customer and in such situations the due date is considered as a decision variable. The problem in which the due date is a decision variable is called the *due date assignment and scheduling problem.* A recent, comprehensive survey of results concerning the common due date assignment and scheduling problems can be found in Gordon et al. [108].

The complexity of the earliness/tardiness scheduling problems depends on the considered objective function. From the point of view of practical applications the most important cost functions are linear and quadratic functions. They are also most exhaustively examined in the literature. In Section 3.1 we discuss problem complexity and algorithms for linear cost functions, and in Section 3.2 for quadratic cost functions.

3.1 Linear cost functions

Linear cost functions are important both from the practical and theoretical points of view. In practice, it is natural to represent earliness (or tardiness) costs as proportional to the earliness (tardiness) length. Moreover, for several problems with linear cost functions optimal schedules can be found in polynomial time.

In the case of a common due date and linear cost functions, the objective is to minimize function:

$$\sum_{i=1}^{n} (\alpha_i \max\{0, d - C_i\} + \beta_i \max\{0, C_i - d\}) \tag{3.1}$$

where coefficients α_i, β_i, $i = 1, \ldots, n$ are the unit earliness and tardiness costs, respectively. The single machine earliness/tardiness scheduling problems with a common due date and linear cost functions are classified depending on the coefficients α_i and β_i. The problem with $\alpha_i = \beta_i = 1$, $i = 1, \ldots, n$, called the mean absolute deviation (MAD) problem, is presented in Section 3.1.1. If $\alpha_i = \alpha, \beta_i = \beta$, $i = 1, \ldots, n$, the problem is called weighted sum of absolute deviations (WSAD). It is examined is Section 3.1.2. The problem with symmetric weights, $\alpha_i = \beta_i$, $i = 1, \ldots, n$, is considered in Section 3.1.3. The formulation with arbitrary weights, called the total weighted earliness and tardiness problem (TWET), is discussed in Section 3.1.4. In Section 3.1.5 we present the problem where the due date is a decision variable, and in Section 3.1.6 the problem with controllable processing times. Problems with resource dependent ready times are presented in Section 3.1.7. Finally, in Section 3.1.8 we consider the problem where zero cost is incurred by tasks completed in a given time window.

3.1.1 Mean Absolute Deviation

In this section we consider the earliness/tardiness problem with identical linear earliness and tardiness costs, i.e. $\alpha_i = \beta_i = 1$, $i = 1, \ldots, n$. Let us consider a set of n tasks with a common due date d, all available for processing at time 0. After reduction, the objective function may be formulated as a sum of absolute deviations of task completion times from the common due date (3.2).

$$f(S) = \sum_{i=1}^{n} |C_i - d|. \tag{3.2}$$

Since both the functions, mean and total absolute deviation, are equivalent as optimality criteria, this problem is often abbreviated as MAD – mean absolute deviation.

Single machine

Let us start with the single machine problem. The single machine scheduling problem of minimizing the total absolute deviation of task completion times from a common due date was first analyzed by Kanet [140].

If $d \geq \sum_{i=1}^{n} p_i$ then the due date is called *unrestrictive* and the problem is polynomially solvable, as shown in [140].

If preemption is allowed, a solution arbitrarily close to the optimum can be found trivially, as noticed by Kanet [140]. Namely, tasks are performed in an arbitrary order in the time interval $[d - \sum_{i=1}^{n} p_i, d]$. Every task is interrupted when it has an arbitrarily small amount of processing time ε remaining. Only after all tasks have been initiated and interrupted, the remaining amount of work is completed. Since ε can be arbitrarily small, the completion time of each task converges to d, that is for each task i, $i = 1, \ldots, n$,

$$\lim_{\varepsilon \to 0} |C_i - d| = 0 \tag{3.3}$$

Thus, the objective function $f(S)$ converges to 0.

If preemption is not allowed, the following properties hold for an optimal schedule for the problem with an unrestrictive due date ([140]).

Property 3.1 *There is no machine idle time inserted in the optimal schedule.*

Property 3.1 means that $C_{i+1} = C_i + p_{i+1}$. Notice, however, that the machine may be idle before starting the first task, i.e. $C_1 \geq p_1$.

Property 3.2 *The optimal schedule is V-shaped [210], i.e. tasks for which $C_i \leq d$ are scheduled in non-increasing order of processing time and tasks for which $C_i > d$ are scheduled in non-decreasing order of processing times.*

Property 3.2 is easy to prove using the interchange argument ([140]).

Property 3.3 *In an optimal schedule one task completes precisely at the due date d.*

It follows from Property 3.1 that the sequence of tasks and the start time of the first task are sufficient to determine the schedule. In fact, it is sufficient to assign tasks to two sets: set E of tasks completed before or at the due date and set T of tasks completed after the due date. The order of tasks in each set follows from Property 3.2. Finally, by Property 3.3 we can calculate the start time of the first task as $d - \sum_{i \in E} p_i$.

Let us consider sets E and T defined above and assume that the sequences of tasks in sets E and T are ordered as stated in Property 3.2. We denote by p_i^E (p_i^T) the processing time of the i-th task in the ordered set E (set T). Finally, let n_E be the number of tasks in set E and n_T be the number of tasks in set T. Notice that by Property 3.3 task number n_E in set E is completed exactly at the due date. The total earliness is then calculated as follows:

$$\sum_{i \in E} (d - C_i) = \sum_{i=1}^{n_E - 1} \sum_{k=i+1}^{n_E} p_k^E$$
$$= (n_E - 1)p_{n_E}^E + (n_E - 2)p_2^E + \ldots 1p_2^E + 0p_1^E.$$

Correspondingly, the total tardiness is calculated as

$$\sum_{i \in T} (C_i - d) = \sum_{i=1}^{n_T} \sum_{k=1}^{i} p_k^T$$
$$= n_T p_1^T + (n_T - 1)p_2^T + \ldots 1p_{n_T}^T.$$

The objective function is the sum of the earliness and tardiness of all tasks:

$$\sum_{i=1}^{n} |C_i - d| = 0p_1^E + 1p_2^E + 1p_{n_T}^T + \ldots + (n_E - 1)p_{n_E}^E + n_T p_1^T. \quad (3.4)$$

It follows from equation (3.4) that an integer coefficient may be associated with each position in the schedule independent of the task assigned to this position. Thus the scheduling problem reduces to assigning each task to a position in the schedule so that the function (3.4) is minimized. It is easy to see that an optimal assignment is obtained if the longest task is assigned to the position with the smallest coefficient, the second longest task to the position with the second smallest coefficient and so on ([120]). Thus, the longest task should be scheduled first. Moreover, if n is even then $n_E = n_T$ and if n is odd then $n_E = n_T + 1$.

Concluding, we can formulate the following property.

Property 3.4 *In an optimal schedule the number of tasks scheduled before the due date equals* $n_E = \lceil n/2 \rceil$.

Equation (3.4) also shows that the MAD problem reduces to the problem of scheduling $n - 1$ tasks (the longest task is dropped) on two identical parallel machines to minimize the mean flow time (see [140, 164]).

The first algorithm solving the MAD problem was proposed by Kanet [140]. We present this algorithm below. Let J denote the set of tasks to be scheduled.

Algorithm 3.1 (Kanet [140]).

1. Set $E = \emptyset$ and $T = \emptyset$.
2. Remove task k such that $p_k = \max\{p_i\}$ from set J.
3. Insert task k into the last position in sequence E.
4. If $J \neq \emptyset$ then remove task k such that $p_k = \max\{p_i\}$ from set J.
5. Insert task k into the first position in sequence T.
6. If $J \neq \emptyset$ then go to step 2.
7. Concatenate sequences E and T.

The final schedule is defined by concatenation (E, T) and contains no idle time. The schedule is obtained by starting the first task in T at time d (or in other words, starting the schedule at time $d - \sum_{i \in E} p_i$). The complexity of the algorithm is $O(nlogn)$. The following example illustrates the idea of the algorithm.

Example 3.2. Let us consider $n = 6$ tasks with processing times $p_1 = 2$, $p_2 = 10$, $p_3 = 4$, $p_4 = 5$, $p_5 = 6$, $p_6 = 12$, and a common due date $d = 90$. The sequences obtained using Algorithm 3.1 are $E = (6, 5, 3)$ and $T = (1, 4, 2)$, and the schedule starts at time $90 - 12 - 6 - 4 = 58$, as shown in Figure 3.1.

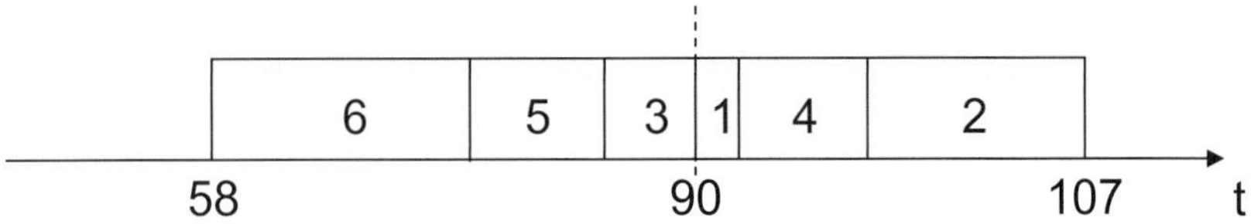

Fig. 3.1. A schedule of tasks from Example 3.2, obtained by Algorithm 3.1

It may be observed that multiple optimal schedules may exist. Bagchi et al. show in [13] that if any two processing times are different then the total number of optimal solutions is 2^k, where $k = \lfloor (n-1)/2 \rfloor$, and they provide a procedure for finding all the optimal solutions. If there are ties, the number of solutions is obviously larger. The algorithm proposed by Bagchi et al. in [13] to find all optimal sequences for the MAD problem is presented below. We denote by $S_i = (E_i, T_i)$ a partial schedule for a subset of tasks $\{1, \ldots, i\}$, and by W_i the set of all partial schedules S_i.

Algorithm 3.3 (Bagchi et al. [13]).

1. If $n = 1$ then go to step 10.
2. Index the tasks according to non-increasing processing times.
3. Set $E_i = \emptyset$, $T_i = \emptyset$, $W_i = \emptyset$, for $i = 1, \ldots, n$.
4. Set $i = 1$, $E_1 = \{1\}$, $W_1 = \{(E_1, T_1)\}$.
5. Remove a partial schedule (E_i, T_i) from W_i. If $||E_i|| \geq ||T_i|| + 1$ then go to step 6, else go to step 7.
6. Construct T_{i+1} by adding task $i+1$ at the begining of T_i. Set $E_{i+1} = E_i$, $S_{i+1} = (E_{i+1}, T_{i+1})$ and $W_{i+1} = W_{i+1} \cup \{S_{i+1}\}$. If $||E_i|| > ||T_i|| + 1$ then go to step 8.
7. Construct E_{i+1} by adding task $i+1$ at the end of E_i. Set $T_{i+1} = T_i$, $S_{i+1} = (E_{i+1}, T_{i+1})$ and $W_{i+1} = W_{i+1} \cup \{S_{i+1}\}$.
8. If $W_i \neq \emptyset$ then go to step 5.
9. Set $i = i + 1$. If $i \neq n$ then go to step 5, else all schedules in W_n are optimal. Stop.
10. The optimal schedule is $(1, \emptyset)$. Stop

Let us illustrate the algorithm with an example.

Example 3.4. Let us consider $n = 5$ tasks with processing times $p_1 = 10$, $p_2 = 4$, $p_3 = 5$, $p_4 = 6$, $p_5 = 12$, and a common due date $d = 90$. The sequences obtained using Algorithm 3.3 are shown in Figure 3.2.

Sometimes it may be necessary to choose one from among many optimal schedules. Bagchi et al. ([13]) consider minimizing the total

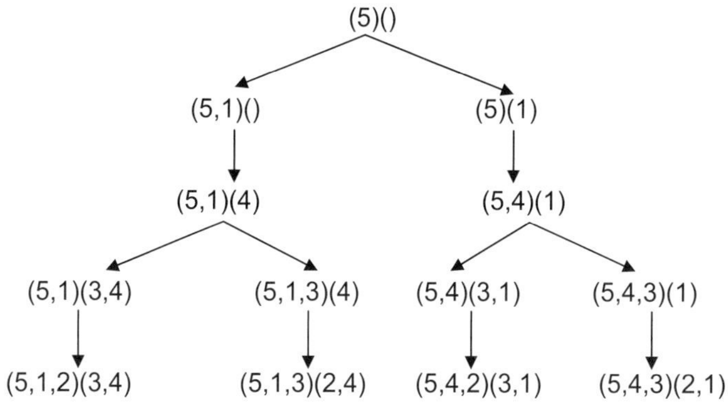

Fig. 3.2. Optimal schedules of tasks from Example 3.4, obtained by Algorithm 3.3

processing time in set E as a secondary criterion. The solution of the extended problem yields the smallest unrestrictive value of d. Notice that the Kanet algorithm finds the solution with the minimum total processing time of tasks in set E.

The schedule obtained by Algorithm 3.1 is optimal if $d \geq \sum_{i \in E} p_i$. If $d < \sum_{i \in E} p_i$ the common due date is called *restrictive*. In this case Properties 3.1 and 3.2 still hold, while Properties 3.3, and 3.4 do not hold. An optimal solution may contain a *straddling* task, i.e. one that starts before d and completes after d. However, the following properties can be proved in this case [118].

Property 3.5 *For any instance of the problem there exists an optimal schedule either starting at time zero or such that one task, k, completes at the due date, i.e. $C_k = d$.*

Property 3.6 *For any instance of the problem, in each optimal schedule, if task k completes at time d, i.e. $C_k{=}d$, then $p_k \leq \max\{\min_{i \in E} \{p_i\}, \min_{i \in T} \{p_i\}\}$.*

Hall et al. in [118] and Hoogeven and van de Velde in [123] prove that the decision version of the MAD problem with restrictive due date is NP-complete. The reduction is from the even-odd partition problem, which is known to be NP-complete in the ordinary sense (see [104]). In fact, in some cases, the problem can be solved in polynomial time even if $d < \sum_{i=1}^{n} p_i$, as stated in Theorem 3.5.

Theorem 3.5 ([118]). *If there exists $k < \lceil n/2 \rceil$, satisfying $\sum_{i=n-k+1}^{n} p_i < d$, and $p_1, \ldots, p_{n-k} > 2d$, then a schedule $S^*{=}(E,T)$ starting at time*

$d - \sum_{i \in E} p_i$, where $E = \lceil n/2 \rceil + 1, \ldots, n$ and $T = 1, 2, \ldots, \lceil n/2 \rceil$ is optimal.

Pseudopolynomial time algorithms for the MAD problem with restrictive due date are presented in [118, 123, 83], and [244]. Thus the problem is NP-hard in the ordinary sense. The computational complexity of the algorithm proposed by Ventura and Weng in [244] is $O[n(d + p_{max})]$. This algorithm is simpler than the algorithm of complexity $O(n \sum p_i)$ presented by Hall et al. in [118]. The procedure developed by Hoogeven and van de Velde requires $O(n^2 d)$ computational time. Below we present the algorithm proposed in [244].

Let us assume that the tasks are ordered according to non-decreasing processing times. The procedure considers tasks starting from the shortest one, i.e. task number 1. Let us denote by $h_k^*(s)$ the minimum cost of scheduling the first k tasks, given that the schedule starts at time s. If $s \leq d$ then at the stage of scheduling task $k+1$ two decisions have to be considered, task $k+1$ can be scheduled as the first task, starting at time s or as the last one, then starting at time $s + \sum_{i=1}^{k} p_i$. The first decision leads to the partial schedule presented in Figure 3.3. The cost of the partial schedule is given by (3.5).

$$h_{k+1}(s, 1) = h_k^*(s + p_{k+1}) + |d - s - p_{k+1}|. \tag{3.5}$$

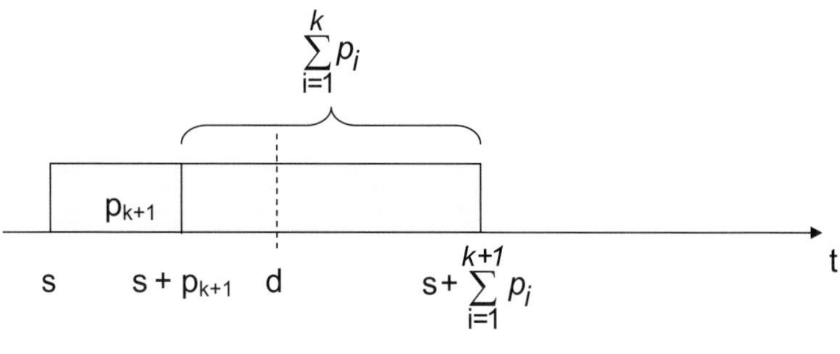

Fig. 3.3. A partial schedule with task $k+1$ starting at time s

The second decision leads to the partial schedule presented in Figure 3.4. The cost of the partial schedule is given by (3.6).

$$h_{k+1}(s, 2) = h_k^*(s) + |s + \sum_{i}^{k+1} p_i - d|. \tag{3.6}$$

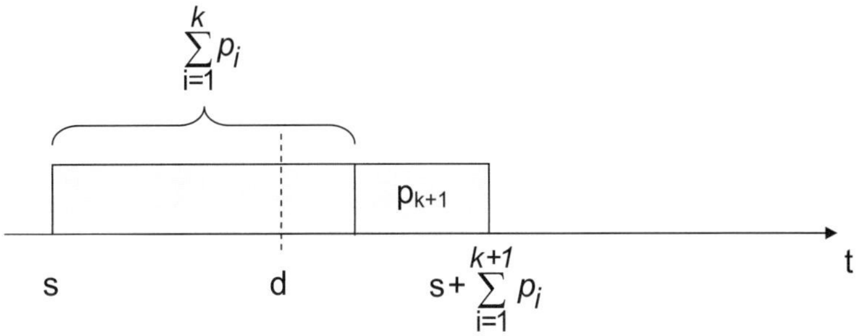

Fig. 3.4. A partial schedule with task $k+1$ starting at time $s + \sum_{i=1}^{k} p_i$

If $s > d$ then all tasks belong to set T, so task $k+1$ has to be scheduled as the last one and the cost of this partial schedule can be calculated according to formula 3.6.

We obtain the following recurrence relation.

$$h^*_{k+1}(s) = \begin{cases} \min\left\{h^*_k(s+p_{k+1}) + |d-s-p_{k+1}|, \right. \\ \qquad \left. h^*_k(s) + |s + \sum_{i=1}^{k+1} p_i - d|\right\} & \text{if } d-s \geq 0 \\ \\ h^*_k(s) + s + \sum_{i=1}^{k+1} p_i - d & \text{otherwise} \end{cases}$$

The dynamic programming procedure is initialized with $h^*_0(s) = 0$ for all s. The minimum cost is found for $\min_{0 \leq s \leq d} h^*_n(s)$.

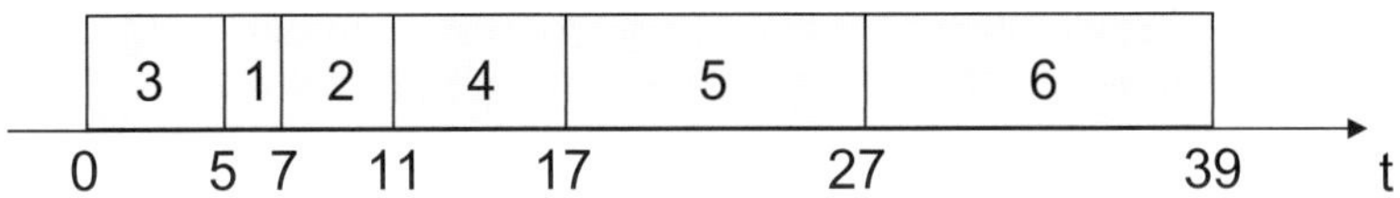

Fig. 3.5. A schedule of tasks from Example 3.6 obtained by the dynamic programming algorithm

Example 3.6. Let us consider $n = 6$ tasks with processing times $p_1 = 2$, $p_2 = 4$, $p_3 = 5$, $p_4 = 6$, $p_5 = 10$, $p_6 = 12$, and a common due date $d = 5$. We start with scheduling the first task. The cost of scheduling a single task depends only on the schedule start time s. The values $h_k(s,1), h_k(s,2)$ and $h^*_k(s)$ for $s = 1, \ldots, 10$, are presented in Table

3.1. It is easy to calculate the values $h_k^*(s)$ for $s > d = 5$ as follows: $h_k^*(s+1) = h_k^*(s)+k$. The corresponding schedule is presented in Figure 3.5.

Two heuristic algorithms have been proposed to solve the MAD problem with a restrictive due date. The first one is developed by Sundararaghavan and Ahmed in [227] and the second one, called *even-odd heuristic*, is proposed by Hoogeven et al. in [124].

The first algorithm is a greedy heuristic. Initially, it is assumed that the available processing time for tasks in set E equals d and the available processing time for tasks in set T equals $\sum_{i=1}^{n} p_i - d$. Since we consider a restrictive due date, the latter value is positive. Tasks ordered according to the LPT rule (longest processing time first) are assigned one by one either to set E or T, depending on the amount of processing time left. The algorithm is presented below.

Algorithm 3.7 (Sundararaghavan and Ahmed [227]).

1. Order tasks in non-increasing order of processing times. Set $k = n$, $R = \sum_{i=1}^{n} p_i - d$, $L = d$, $E = \emptyset$ and $T = \emptyset$.
2. If $R \geq L$ then assign task k to the first posision in set T and set $R = R - p_k$.
3. If $R < L$ then assign task k to the last posision in set E and set $L = L - p_k$.
4. If $k > 1$ then set $k = k - 1$ and go to step 2.
5. Concatenate sequences E and T and start the schedule at time zero.

We illustrate the heuristic with the following example.

Example 3.8. Let us consider $n = 8$ tasks with processing times $p_1 = 2$, $p_2 = 3$, $p_3 = 5$, $p_4 = 6$, $p_5 = 8$, $p_6 = 9$, $p_7 = 12$ and $p_8 = 14$, and let $d = 15$. The values of R and L in consecutive iterations, as well as the sets E and T, are presented in Table 3.2. The schedule obtained by the algorithm is presented in Figure 3.6.

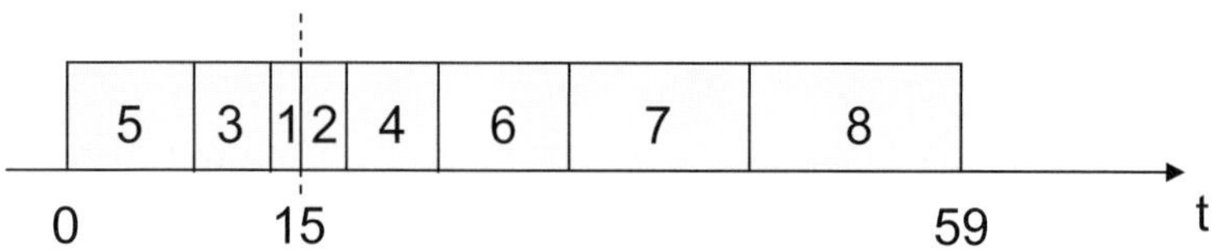

Fig. 3.6. A schedule of tasks from Example 3.8, obtained by Algorithm 3.7

Table 3.1. Solution of the problem from Example 3.6

	Task 1	Task 2			Task 3			Task 4			Task 5			Task 6		
s	$h_1^*(s)$	$h_2(s,1)$	$h_2(s,2)$	$h_2^*(s)$	$h_3(s,1)$	$h_3(s,2)$	$h_3^*(s)$	$h_4(s,1)$	$h_4(s,2)$	$h_4^*(s)$	$h_5(s,1)$	$h_5(s,2)$	$h_5^*(s)$	$h_6(s,1)$	$h_6(s,2)$	$h_6^*(s)$
0	3	2	4	2	8	8	**8**	23	20	**20**	61	42	**42**	105	76	**76**
1	2	2	4	2	11	9	9	25	22	22	64	45	45	109	80	80
2	1	4	4	4	14	12	12	29	26	26	67	50	50	113	86	86
3	0	6	4	4	17	13	13	33	28	28	70	53	53	117	90	90
4	1	8	6	6	20	16	16	37	32	32	73	58	58	121	96	96
5	**2**	10	8	**8**	23	19	19	41	36	36	76	63	63	125	102	102
6	3	12	10	10	26	22	22	45	40	40	81	68	68	129	108	108
7	4	14	12	12	29	25	25	49	44	44	86	73	73	133	114	114
8	5	16	14	14	32	28	28	53	48	48	91	78	78	139	120	120
9	6	18	16	16	35	31	31	57	52	52	96	83	83	145	126	126
10	7	20	18	18	38	34	34	61	56	56	101	88	88	151	132	132

Table 3.2. Solution to the problem from Example 3.8

Iteration	R	L	T	E
0	44	15	8	–
1	30	15	8,7	–
2	18	15	8,7,6	–
3	9	15	8,7,6	5
4	9	7	8,7,6,4	5
5	3	7	8,7,6,4	5,3
6	3	2	8,7,6,4,2	5,3
7	0	2	8,7,6,4,2	5,3,1

Let us now present the even-odd heuristic. Hoogeven et al. [124] formulate the MAD problem with a restrictive due date as the problem of minimizing the total absolute deviation with the constraint that the amount of work scheduled before the due date may not exceed d. The Lagrangian relaxation of this problem is formulated as follows:

$$\text{minimize} \quad L(\lambda) = \sum_{i=1}^{n} (e_i + t_i) + \lambda(W - d) \tag{3.7}$$

where W is the amount of work (total processing time) that is processed up to time d and λ is a Lagrangian multiplier. Obviously, the optimal solution $L^*(\lambda)$ of problem (3.7) is a lower bound to the original MAD problem with the restrictive due date d.

Hoogeven et al. show in [124] that problem (3.7) can be solved in polynomial time using Algorithm 3.9, which is a slight modification of the algorithm given by Emmons in [95].

Algorithm 3.9 (Emmons [95]).

1. Assign position i in the sequence a weight w_i, $i = 1, \ldots, n$, calculated as follows:
$$w_i = \begin{cases} i - 1 + \lambda \text{ if } i \text{ occurs before } d \\ n - i + 1 \ \text{ if } i \text{ occurs after } d \end{cases} \tag{3.8}$$
2. Index the tasks in the non-increasing order of processing times.
3. Index the positions in the non-decreasing order of weights.
4. Assign tasks i to position i, $i = 1, \ldots, n$.

The problem of finding the value λ^* that maximizes the lower bound is called the Lagrangian dual problem. Theorem 3.10 states that this problem can be solved in polynomial time.

Theorem 3.10 ([124]). *The optimal value λ^*, i.e. the value that maximizes the Lagrangian lower bound, is equal to the index λ for which*

$$\sum_{i=0}^{\lfloor (n-\lambda)/2 \rfloor} p_{\lambda+2i} \geq \sum_{i=0}^{\lfloor (n-\lambda-1)/2 \rfloor} p_{\lambda+1+2i}. \tag{3.9}$$

If no such index exists, then $\lambda^* = 0$.

If there exists a solution of the Lagrangian dual problem, such that the first task starts at time zero, then it is an optimal solution of the MAD problem.

In order to find a better lower bound we solve a modified Lagrangian problem, formulated as follows:

$$\text{minimize} \;\; L_m(\lambda^*) = \sum_{i=1}^{n} |C_i - d| + \lambda^*(W - d) + |W - d|. \tag{3.10}$$

Theorem 3.11 [124] characterizes the solution of the modified Lagrangian problem.

Theorem 3.11 ([124]). *The modified Lagrangian problem is solved by a schedule from among the optimal schedules for the Lagrangian dual problem that has the minimal value of $|W - d|$.*

Any instance of the problem of minimizing $|W-d|$ is transformed to a considerably smaller instance of the knapsack problem (or the subset sum problem). The knapsack problem is formulated as follows.

Problem 3.12 (Knapsack problem). Consider a set of n items with weights w_i and values p_i, and an integer B. Find a subset of items with maximum total value and total weight not exceeding B, i.e.

$$\begin{aligned} \text{maximize} \;\; & \sum_{i=n}^{n} p_i x_i \\ \text{subject to} \;\; & \sum_{i=1}^{n} w_i x_i \leq B \\ & x_i \in \{0, 1\}, i = 1, \ldots n. \end{aligned}$$

The knapsack problem is NP-hard in the ordinary sense (see [103]).

An optimal solution of a given instance of the knapsack problem can be found by a dynamic programming algorithm in $O(nB)$ time, where B is the size of the subset and n is the number of elements. In Algorithm 3.14 a heuristic proposed by Johnson in [130] is used to find an approximate solution of the subset sum problem in $O(n)$ time after sorting. Johnson heuristic adds elements to the subset in the non-decreasing order of their weights as long as the sum of weights in the subset is less than or equal to B. Given is a set of n elements ordered accaording to non-decreasing weights w_i, $i = 1, \ldots, n$. The Johnson algorithm is presented below.

Algorithm 3.13 (Johnson [130]).

1. Set $i = 1$, $A = \emptyset$.
2. If $w_i \le B$ then set $A = A \cup \{i\}$ and set $B = B - w_i$.
3. If $i < n$ then set $i = i + 1$ and go to step 2, else stop.

For each $\lambda = 1, \ldots, n$, we denote by W_λ^{min} and W_λ^{max} the amount of work completed before the due date in an optimal solution of the Lagrangian problem with minimum, respectively maximum, value W. Let the tasks be indexed according to the non-decreasing processing times. We define $w_i = p_{i+1} - p_i$ and apply Algorithm 3.13 to find a subset of elements with the target sum equal to $B = d - W_\lambda^{min}$. The solution is the set A_1 with deviation from the target sum denoted as B_1. Moreover, we apply Algorithm 3.13 to solve the subset sum problem for the target sum equal to $\sum_{i=1}^{m} w_i - B$. We denote by A_2 the complement of the obtained subset and by B_2 the resulting deviation of A_2 from the target sum B. If both $B_1 > 0$ and $B_2 > 0$, then a feasible schedule is constructed using Algorithm 3.14.

Algorithm 3.14 (Transform ([124])).

1. Consider the solution of the Lagrangian dual problem with minimum workload scheduled before the due date. Interchange the tasks that correspond to the element w_i, $i = 1, \ldots, n$ in set A_1. Each interchange increases W by w_i.
2. Shift the schedule obtained in step 1 to the left, so that the first task starts at time zero or the number of early tasks exceeds the number of late tasks by 2. Rearrange the tasks to receive a V-shape schedule.
3. Consider the solution of the dual Lagrangian problem with maximum workload before the due date. Interchange tasks that correspond to element w_i in set A_2. Each interchange decreases W by w_i.
4. Shift the schedule to the right by B_2, so that the first task starts at time zero, and rearrange the tasks to receive a V-shape schedule.

If some task starts before the due date and completes after the due date, shift the schedule to the right, so that this task starts at the due date. Again rearrange the tasks to receive a V-shape schedule.

Based on the solution of the modified Lagrangian problem, Hoogeven et al. [124] propose the following heuristic algorithm, running in $O(nlogn)$ time.

Algorithm 3.15 (Even-odd heuristic Hoogeven et al. [124]).

1. Solve the Lagrangian dual problem and apply Algorithm 3.13 to the corresponding instance of the subset sum problem with $B = d - W_{\lambda^*}^{min}$. Let B_1 denote the gap for the approximation from above and B_2 the gap from below.
2. If $B_1 \leq B_2$ then apply Algorithm 3.14 and go to step 6.
3. Set $Q = \{w_i : w_i \geq B_1\}$. If $Q \neq \{w_1\}$ then apply Algorithm 3.14 and go to step 6.
4. If $p_1 > d$ then apply Algorithm 3.14 and solve the Lagrangian dual problem under the condition that the shortest task and all the remaining tasks are assigned to the last positions, then go to step 6.
5. Solve the Lagrangian dual problem under the condition that the shortest task and all the remaining tasks are assigned to positions after d, and solve the Lagrangian dual problem with the shortest task assigned to a position before d. Apply Algorithm 3.13 and Algorithm 3.14 to all these solutions.
6. Choose the schedule with the minimum cost.

This algorithm has the performance guarantee 4/3 (see [124]) and this bound can be approximated arbitrarily closely. We illustrate the algorithm with an example.

Example 3.16. Let us consider $n = 8$ tasks with processing times $p_1 = 2$, $p_2 = 3$, $p_3 = 5$, $p_4 = 6$, $p_5 = 8$, $p_6 = 9$, $p_7 = 12$, and $p_8 = 14$, and let $d = 16$. We first solve the dual Lagrangian problem with minimum workload scheduled before the due date. The solution is obtained for $\lambda = 3$ and the sequence of tasks is the following: $5, 3, 1, 2, 4, 6, 7, 8$. Applying Algorithm 3.13 we obtain set $A_1 = 1$ with $B_1 = 0$. Now, we apply Algorithm 3.14 and interchange tasks 1 and 2. The resulting sequence is the following: $5, 3, 1, 2, 4, 6, 7, 8$, and since the schedule starts at time zero, it is optimal.

Branch and bound algorithms for the restricted MAD problem are proposed by Bagchi et al. [13] and Szwarc in [231]. None of these algorithms solved instances with more than 25 tasks. Hoogeven et al. in

[124] propose a branch and bound algorithm with Lagrangian lower and upper bounds. The branch and bound algorithm runs in three phases. The Lagrangian dual problem is solved in the first phase. If $\lambda^* = 0$ then the obtained sequence is an optimal solution and no branching is necessary. Otherwise, the upper bound is determined by the better one of the two solutions found respectively by Algorithm 3.7 and Algorithm 3.15. If the lower bound is not equal to the upper bound then an optimal solution of the subset sum problem is found using the dynamic programming procedure. If the bounds still differ the branch and bound procedure is applied.

The branching scheme takes advantage of the V-shape of optimal schedules. It is assumed that tasks are indexed according to the non-increasing processing times. A node at level $k, k = 1, \ldots, n$, defines a path in the branching tree from the root to this node. If task i, $k \leq i \leq n$, is assigned to a node at level k, then the partial schedule consisting of i tasks is defined in the following way: tasks assigned to nodes on the path from the root to node k are ordered according to the LPT rule and the first one starts at time zero. If $i > k$ then all the tasks with $1 \leq j \leq i$ not assigned to any node on the path are ordered according to the SPT rule (shortest processing time first) and the last one is completed at time $\sum_{i=1}^{n} p_i$. Each node at level k has at most $n - k$ descendants corresponding to the unscheduled tasks. Depth-first search is applied with an active node search. At each level one node is chosen to branch from. It is the node whose task has the smallest remaining index. In order to avoid duplication of solutions, a node at level k corresponding to some task i can be discarded if another node at the same level corresponding to some task j with $p_i = p_j$ has already been considered. At each node the lower bound $L(\lambda^*)$ is calculated. Neither the solution of the modified Lagrangian problem nor any additional upper bounds are computed.

The branch and bound algorithm described above finds optimal solutions for instances with up to 1000 tasks within 1 second on a Compaq-386 personal computer ([124]).

Identical parallel machines

Let us now examine the problem of minimizing the mean absolute deviation in the system of identical parallel machines. Hall in [115] and Sundararaghavan and Ahmed in [227] prove the following properties of optimal schedules for this problem:

Property 3.7 *There is no inserted idle time on any machine.*

Property 3.8 *On each machine the optimal schedule is V-shaped.*

Property 3.9 *On each machine one task completes at time d.*

Property 3.10 *The number of tasks assigned to each of the m machines is either $\lfloor n/m \rfloor$ or $\lceil n/m \rceil$. The task number $\lceil n_j/2 \rceil$ completes at time d on machine j, where n_j is the number of tasks assigned to machine j.*

It follows from Properties 3.7 - 3.10 that if the optimal assignment of tasks to machines is known, then an optimal schedule on each machine can be found using Algorithm 3.1. An optimal assignment of tasks to machines can be found in polynomial time using Algorithm 3.17. Concluding, the complexity of minimization of the mean absolute deviation from the common unrestrictive due date in the system of parallel identical machines is $O(nlogn)$. Let us assume that tasks are indexed according to the LPT rule, i.e. $p_1 \geq p_2 \geq \ldots \geq p_n$.

Algorithm 3.17 (Sundararaghavan and Ahmed [227]).

1. Assign m tasks $1, \ldots, m$, one to each machine. Set $n = n - m$.
2. If $2m \leq n$ then assign the next $2m$ tasks to machines, two to each machine, set $n = n - 2m$ and go to step 2.
3. If $m < n < 2m$ then assign the next m tasks one to each machine and set $n = n - m$.
4. If $0 \leq n \leq m$ then assign the tasks to the machines, so that each machine receives at most one task.
5. Find an optimal solution on each machine using Algorithm 3.1.

We illustrate Algorithm 3.17 with the following example.

Example 3.18. Let us consider $n = 14$ tasks with processing times $p_1 = 37$, $p_2 = 35$, $p_3 = 32$, $p_4 = 31$, $p_5 = 27$, $p_6 = 23$, $p_7 = 21$, $p_8 = 19$, $p_9 = 16$, $p_{10} = 12$, $p_{11} = 11$, $p_{12} = 7$, $p_{13} = 5$, $p_{14} = 3$, to be scheduled on three machines and let the common due date $d = 100$. Let M_j denote the set of tasks assigned to machine j. The following assignment of tasks to machines may be obtained using Algorithm 3.17: $M_1 = \{1, 4, 5, 10, 13\}$, $M_2 = \{2, 6, 7, 11, 14\}$, and $M_3 = \{3, 8, 9, 12\}$. Next, three single-machine MAD problems with $d = 100$ have to be solved for the following sets of tasks:

J_1: $n = 5$ tasks with processing times $37, 31, 27, 12$, and 5, respectively.
J_2: $n = 5$ tasks with processing times $35, 23, 21, 11$, and 3, respectively.
J_3: $n = 4$ tasks with processing times $32, 19, 16$, and 7, respectively.

An optimal schedule for the three machine problem given in Example 3.18 is presented in Figure 3.7.

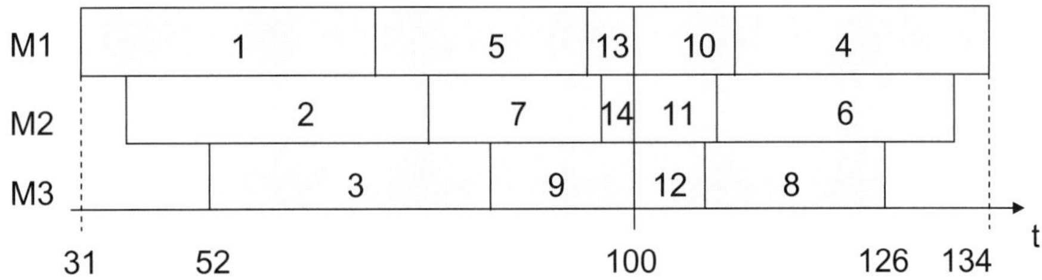

Fig. 3.7. An optimal schedule of tasks from Example 3.18

Uniform parallel machines

The problem of scheduling tasks on parallel uniform machines with unrestrictive due date can be also solved in polynomial time. An $O(nlogn)$ algorithm is proposed by Emmons in [95]. The idea is similar to the one applied for the system of identical parallel machines and consists in assigning weights to positions of tasks in a schedule. Processing times of tasks on different uniform machines are different but proportional. Let us denote by η_j, $0 < \eta_j < \infty$, the processing speed of machine j, $j = 1, \ldots, m$. Let machine 1 be the standard machine on which the task processing times p_i, $i = 1, \ldots, n$, are defined. If the processing time of a task on the standard machine is p_i then its processing time on machine j is equal to p_i/η_j. We can now modify the weights of the objective function (3.4) to obtain:

$$\sum_{i=1}^{n} |C_i - d| = \sum_{k=1}^{m} \Big(\sum_{j \in E_k} \frac{(j-1)}{\eta_k} p_{jk} + \sum_{j \in T_k} \frac{j}{\eta_k} p_{jk} \Big). \tag{3.11}$$

where E_k (T_k) is the set of early (tardy) tasks scheduled on machine k, and p_{jk} is the processing time of the task which has $j-1$ predecessors (successors) on the k-th machine. The tasks are arranged in the order of non-increasing processing times and each task is assigned to the position with the currently smallest weight. The algorithm is presented below in a more formal way. It is assumed that tasks are indexed in the non-decreasing order of their processing times.

Algorithm 3.19 (Emmons [95]).

1. Schedule m longest tasks, one on each machine. Set $i = m$.
2. Set $w_{ek} = 1/\eta_k$ and $w_{tk} = 1/\eta_k$, $k = 1, \ldots, m$.
3. Schedule task $n - i$ in the position with the smallest weight w_{xl},

$$w_{xl} = \min_{k=1,\ldots,m} \{w_{ek}, w_{tk}\}$$

and set $w_{xl} = w_{xl} + 1/\eta_l$.
4. If $i < n$ then Set $i = i + 1$ and go to step 3, else stop.

We use Algorithm 3.19 to find an optimal schedule for the following example.

Example 3.20. Let us consider an instance with $n = 25$ tasks and $m = 3$ machines. The processing times of the tasks are the following: $p_1 = 1$, $p_2 = 2$, $p_3 = 4$, $p_4 = 5$, $p_5 = 7$, $p_6 = 8$, $p_7 = 10$, $p_8 = 11$, $p_9 = 12$, $p_{10} = 14$, $p_{11} = 15$, $p_{12} = 17$, $p_{13} = 18$, $p_{14} = 21$, $p_{15} = 23$, $p_{16} = 27$, $p_{17} = 32$, $p_{18} = 33$, $p_{19} = 35$, $p_{20} = 36$, $p_{21} = 38$, $p_{22} = 42$, $p_{23} = 43$, $p_{24} = 45$, and $p_{25} = 47$, and the due date is $d = 1000$. The speeds of particular machines are: $\eta_1 = 1$, $\eta_2 = 1/2$ and $\eta_3 = 1/3$. The weights calculated for particular positions on the three machines are presented in Table 3.3. Three longest tasks are assigned to positions with weights equal to zero. The 25 positions with minimum weights are displayed in boldface. The corresponding schedule is presented in Figure 3.8.

Table 3.3. Weights of particular positions on machines in Example 3.20

		number of predecessors/successors									
Machine	set	1	2	3	4	5	6	7	8	9	10
Machine 1	E	**0**	**1**	**2**	**3**	**4**	**5**	**6**	7	8	9
	T	**1**	**2**	**3**	**4**	**5**	**6**	7	8	9	10
Machine 2	E	**0**	**2**	**4**	**6**	8	10	12	14	16	18
	T	**2**	**4**	**6**	8	10	12	14	16	18	20
Machine 3	E	**0**	**3**	**6**	9	12	15	18	21	24	27
	T	**3**	**6**	9	12	15	18	21	24	27	30

Flow shop

The minimization of the mean absolute deviation in a flow shop system is considered by Sung and Min in [228]. They consider a two-machine flow shop with the objective defined by formula (3.12)

$$\text{minimize} \sum_{i=1}^{n} |C_{i2} - d| \tag{3.12}$$

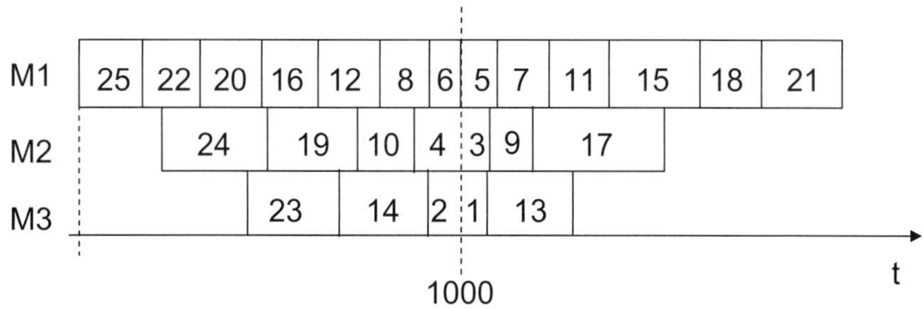

Fig. 3.8. An optimal schedule of tasks from Example 3.20, obtained using Algorithm 3.19

where C_{i2} is the completion time of task i, $i = 1, \ldots, n$, on the second machine. Moreover, it is assumed that the common due date d is larger than the total processing time on the first machine. At least one machine is a batch processing machine, where batch processing is understood as processing all tasks belonging to the same batch simultaneously. Given are the processing time of a batch, p_b, and the maximum number of tasks assigned to one batch c. The following system configurations are considered.

C1 The first machine processes each task separately, while the second machine is a batch processing machine.
C2 Both machines process tasks in batches.
C3 The first machine is a batch processing machine and the second one processes each task separately.

The first two cases are polynomially solvable, while the third one is NP-hard. Below we present the polynomial time algorithms for configurations [C1] and [C2] and a pseudopolynomial dynamic programming algorithm for configuration [C3].

Let us first consider the system configuration [C1] and denote the processing time of task i processed on the non-batching machine by p_i. Let us denote by m the smallest number of batches that can be created in the schedule, $m = \lceil n/c \rceil$.

It is optimal to schedule tasks on the first machine in the SPT order, without idle time inserted. Since the number of batches should be minimized, at most $m = \lceil n/c \rceil$ batches are created, where at most two of them contain less than c tasks. If the due date is long enough, an optimal schedule on the second machine may be easily constructed by scheduling $k = \lceil m/2 \rceil$ batches before the due date and the remaining ones starting at the due date. If m is even, then the batch containing less than c tasks is scheduled last, otherwise it is scheduled as the first batch. However, it may happen that a schedule obtained in this way collides with the schedule on the first machine, meaning that a batch containing some task i starts processing on the second machine before task i is completed on the first machine. Such schedule is obviously infeasible. In this case, the schedule on the second machine is shifted to the right hand side until a feasible optimal schedule is obtained. At most c iterations are required to find this schedule. There may be no batch completed exactly at the due date in such schedule. The following algorithm finds an optimal schedule in $O(nlogn)$ time. We denote by n_E (n_T) the number of early (tardy) tasks, and by n_E^{max} the maximum number of tasks that may be completed before the due date. We calculate n_E^{max} as follows:

$$n_E^{max} = \min_{0 \le j \le k} \{||n_E(j)|| + (k-j)c\}, \tag{3.13}$$

where $n_E(j) = \{i : C_{i1} \le d - (k+1-j)p_b\}$.

Algorithm 3.21 (Sung and Min [228]).

1. Schedule the first operations of each task on the first machine in the SPT order. Calculate the completion times of the operations as follows: $C_{i1} = \sum_{k=1}^{i} p_k$.
2. Calculate the maximum number of early tasks n_E^{max}, according to formula (3.13).
3. If m is even then go to step 4, else go to step 5.
4. If $n_E^{max} \ge mc/2$ then start the schedule on the second machine at time $d - mp_b/2$ and schedule the batch containing less than c tasks as the last one (the total deviation is equal to $\sum_{j=1}^{m} n_j |m/2 - j| p_b$, where n_j is the number of tasks in batch j, j=1, ..., m), else go to step 6.
5. If $n_E^{max} \ge n - (\lceil m/2 \rceil - 1)c$ then start the schedule on the second machine at time $d - \lceil m/2 \rceil p_b$ and schedule the batch containing less than c tasks as the first one (the total deviation is equal to $\sum_{j=1}^{m} n_j |\lceil m/2 \rceil - j| p_b$, where n_j is the number of tasks in batch j, j=1, ..., m), else go to step 6.
6. Set $n_E = n_E^{max}$ and the total deviation $Z^* = \infty$.
7. If $n_E \ge n_T$ then build a minimum number of batches, separately for early and for late tasks. Construct the schedule on the second machine so that only the first batch of early tasks and the last batch of tardy tasks contain less than c tasks and the schedule starts at time $d - kp_b$ (the total deviation is equal to $\sum_{i=1}^{k} n_i(k-i)p_b - \sum_{i=1}^{k'} n_{k+i} i p_b$), and stop.

8. Calculate $k = \lceil n_E/c \rceil$ and $k' = \lceil n_T/c \rceil$ and build a minimum number of batches, separately for early and for late tasks. Construct the schedule so that only the first batch of early tasks and the last batch of tardy tasks contain less than c tasks and the k-th batch completes at the due date. Calculate the maximum possible left shift of the algorithm as follows: $\epsilon = \min\{d - (k - j + 1)p_b - R_j\}$, where $R_j = \max\{C_{i1}$, for all tasks i contained in batch $j\}$. Construct a schedule starting at time $d - kp_b - \epsilon$ with only the first batch of early tasks and the last batch of tardy tasks containing less than c tasks. Set the total absolute deviation to $Z = \sum_{j=1}^{k} n_j((k - j)p_b + \epsilon) + \sum_{j=1}^{k'} n_j(jp_b - \epsilon)$.
9. If $Z^* > Z$ then set $Z^* = Z$, set $n_E = n_E - 1$ and go to step 8, else stop.

Let us illustrate the above algorithm with an example.

Example 3.22. Let us consider an instance of the two machine flow shop scheduling problem with $n = 10$ tasks, $d = 85$, $c = 2$, $p_b = 30$ and task processing times: $p_1 = 1$, $p_2 = 4$, $p_3 = 5$, $p_4 = 5$, $p_5 = 6$, $p_6 = 7$, $p_7 = 8$, $p_8 = 13$, $p_9 = 15$, and $p_{10} = 20$. The task completion times on the first machine are presented in Table 3.4.

Table 3.4. Task completion times on the first machine, Example 3.22

Task i	1	2	3	4	5	6	7	8	9	10
C_{i1}	1	5	10	15	21	28	36	49	64	84

Now we calculate the maximum number of early tasks $n_E^{max} = \min\{4, 5 + 2, 8\} = 4$, and the number of batches $m = \lceil 10/2 \rceil = 5$. Because m is odd we go to step 5. Since $4 = n_E^{max} < n - (\lceil m/2 \rceil - 1)c = 10 - (3 - 1)2 = 6$, we go to step 6 and set $n_E = 4$ and $Z^* = \infty$. Since $4 = n_E < n_T = 3$, we go to step 8 and calculate $k = 2$ and $k' = 3$. The schedule on the second machine, constructed in step 8, starts at time $85 - 60 = 25$ and each batch contains 2 tasks. Now we calculate $\epsilon = \min\{25 - 5, 55 - 15, 85 - 28\} = 20$ and we construct a new schedule starting at time $25 - 20 = 5$. The corresponding total deviation is equal to $Z = 2(30 + 20) + 2(20) + 2(30 - 20) + 2(60 - 20) + 2(90 - 20) = 380$. In the next step we set $Z^* = 380$ and $n_E = 3$ and go to step 7. Now $k = 2$ and $k' = 4$ and $\epsilon = \min\{25 - 1, 55 - 10, 85 - 21\} = 24$. The new schedule starts at time $85 - 60 - 24 = 1$ and has the total deviation equal to $Z = 390$. Since $Z > Z^*$, we stop. The obtained schedule is presented in Figure 3.9.

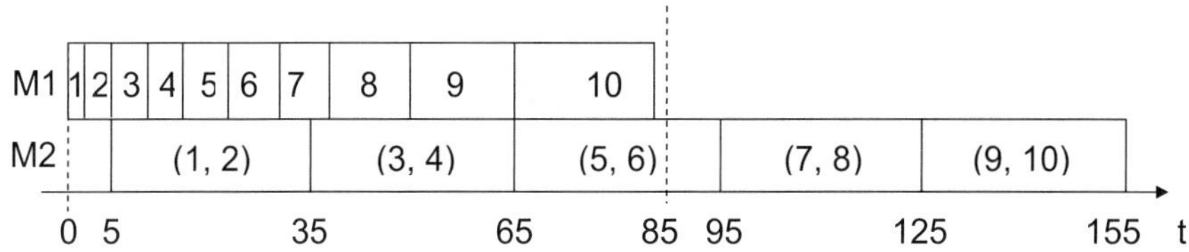

Fig. 3.9. An optimal schedule of tasks from Example 3.22

Let us now consider configuration [C2]. In this system both machines are batch processing machines and we assume that batch processing times and maximum batch sizes are p_{b1} and c_1 on the first machine and p_{b2} and c_2 on the second machine, respectively. In this case the optimal schedule on the first machine starts at time zero with the last batch containing less than c_1 tasks. The optimal schedule on the second machine is now obtained in the same way as in Algorithm 3.21. The only difference is that completion times of particular tasks on the first machine are calculated according to the batch schedule obtained in the first stage of the algorithm, i.e. $C_{j1} = \lceil j/c_1 \rceil p_{b1}$.

Finally, let us consider the MAD problem in a flow shop system with configuration [C3]. In general, for $c = n$ and a short due date, the problem to be solved on the second machine is equivalent to the MAD problem with restrictive due date and so it is NP-hard. A pseudopolynomial time algorithm was proposed in [228] to solve this problem. An optimal schedule on the first machine starts at time zero and only the last batch (if any) contains less than c tasks. The order in which tasks are assigned to each batch depends on the schedule developed for the second machine. If the due date is long, then the schedule on the second machine may be obtained using Algorithm 3.1, otherwise the dynamic programming procedure proposed by Ventura and Weng (see [244]) and presented in Section 3.1.1 is used. The algorithm for the flow shop is formulated below. Let us assume that tasks are indexed according to the SPT rule.

Algorithm 3.23 (Sung and Min [228]).

1. If n is even then set $\Delta = p_n + p_{n-2} + \ldots + p_2$, else set $\Delta = p_n + p_{n-2} + \ldots + p_1$.
2. If $d - t \geq \Delta$ then find an optimal schedule of tasks on the second machine using Algorithm 3.1 and go to step 4.
3. If $d - t < \Delta$, find an optimal schedule on machine $M2$ with $t \leq s \leq d$, using the dynamic programming algorithm (see [244]).

4. Create batches, each one containing c tasks, adding tasks to the batches in the order in which they are scheduled on the second machine. The last batch may contain less than c tasks.

Thus, for configuration [C3] an optimal schedule can be found using pseudopolynomial time Algorithm 3.23.

Concluding, the necessary condition for the MAD problem to be solved in polynomial time is the unrestrictive due date. In the case of the unrestrictive due date polynomial time algorithms are known. In the case of the restrictive due date, even the single machine MAD problem is NP-hard in the ordinary sense. Pseudopolynomial time algorithms are developed for this case. Results and algorithms reported in the literature for the MAD problem are summarized in Table 3.5.

Table 3.5. Complexity of MAD problems

Cost function	Complexity	Algorithms
$1\|d_j = d^{unres}\|\sum(e_i + t_i)$	$O(nlogn)$ [140] [140, 205, 13, 115]	
$1\|d_j = d^{res}\|\sum(e_i + t_i)$	NP-hard [118, 124]	dynamic programming $O(n\sum p_i)$ [118, 244] branch & bound [124]
$Pm\|d_j = d^{unres}\|\sum(e_i + t_i)$	$O(nlogn)$ [227, 115]	
$Qm\|d_j = d^{unres}\|\sum(e_i + t_i)$	$O(nlogn)$ [95]	
$F2[C1]\|d_j = d^{unres}\|\sum(e_i + t_i)$	$O(nlogn)$ [228]	
$F2[C2]\|d_j = d^{unres}\|\sum(e_i + t_i)$	$O(nlogn)$ [228]	
$F2[C3]\|d_j = d^{unres}\|\sum(e_i + t_i)$	NP-hard [228]	dynamic programming $O(n(d + p_{n-1}))$ [228]

3.1.2 Weighted Sum of Absolute Deviations

One of the possible generalizations of the problem of minimizing the absolute deviation from the common due date is the problem where the unit earliness and tardiness costs differ, although they do not depend on tasks, i.e. $\alpha = \alpha_i$, and $\beta = \beta_i$, $i = 1, \ldots, n$. The objective function to be minimized is formulated as follows:

$$\sum_{i=1}^{n} (\alpha e_i + \beta t_i). \tag{3.14}$$

Function (3.14) is called the weighted sum of absolute deviations (WSAD).

Single machine

Let us first consider the problem of scheduling n independent, nonpreemptive tasks on a single machine so as to minimize function (3.14). Similarly as for the MAD problem, the cases of restrictive and unrestrictive due date have to be considered separately. Let us first consider the unrestrictive due date. Properties 3.1 and 3.2 hold for an optimal schedule. Moreover, the task completed exactly at the due date is characterized by Property 3.11.

Property 3.11 *In an optimal schedule, the k-th task in the sequence completes at time d, where k is the smallest integer greater than or equal to $n\beta/(\alpha+\beta)$.*

This statement reduces to Property 3.3 if $\alpha = \beta$.

A polynomial time algorithm for the weighted absolute deviation problem with unrestrictive due date was developed by Bagchi et al. [15]. In order to find the optimal schedule, two ordered sets, E and T, are constructed. We assume that tasks are indexed according to the LPT rule and that $||E||$ and $||T||$ denote the number of tasks in set E and T, respectively. The algorithm is presented below.

Algorithm 3.24 (Bagchi et al. [15]).

1. Set $T = \emptyset$, $E = \emptyset$, and $k = 1$.
2. If $\alpha||E|| < \beta(1 + ||T||)$ then assign task k to the first position in set T, else assign task k to the last position in E.
3. If $k < n$ then set $k = k + 1$ and go to step 2.
4. Concatenate set E, ordered according to non-increasing task processing times, and set T, ordered according to non-decreasing task processing times.

An optimal schedule is obtained by scheduling tasks in the order (E, T) without idle time and starting it at time $d - \sum_{i \in E} p_i$. Multiple optimal solutions to this scheduling problem may exist. Algorithm 3.24 constructs the solution with minimum total processing time of tasks assigned to set E. Thus, similarly to the case with $\alpha = \beta$, the due date d is considered unrestrictive if $d \geq \sum_{i \in E} p_i$.

Let us consider an example illustrating Algorithm 3.24.

Example 3.25. Let us consider an instance with $n = 6$ tasks with processing times $p_1 = 12$, $p_2 = 10$, $p_3 = 6$, $p_4 = 5$, $p_5 = 4$, $p_6 = 2$, $\alpha = 3$, $\beta = 1$, and a common due date $d = 90$. The values $\alpha||E||$ and $\beta(||T|| + 1)$ and the decisions made in consecutive iterations are presented in Table 3.6. The sequences obtained using Algorithm 3.24 are $E = (1, 5)$ and $T = (2, 3, 4, 6)$, and the schedule starts at time $90 - 12 - 4 = 74$ as shown in Figure 3.10.

Table 3.6. Solution of Example 3.25

Iteration	$\alpha\|\|E\|\|$	$\beta(\|\|T\|\| + 1)$	T	E
1	0	1	1	–
2	3	1	1	2
3	3	2	1	2,3
4	3	3	1	2,3,4
5	6	4	1,5	2,3,4
6	6	4	1	2,3,4,6

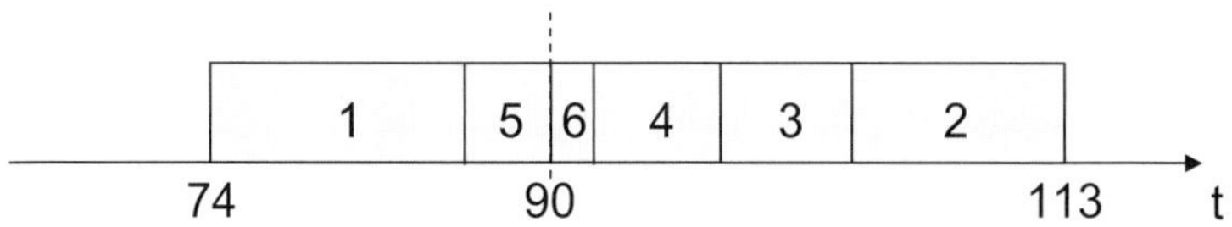

Fig. 3.10. A schedule of tasks from Example 3.25 obtained by the Algorithm 3.24

Minimization of the weighted absolute deviation on a single machine with a restrictive due date is obviously NP-hard. Moreover, Bagchi et al. [15] show that Properties 3.1 and 3.2 also hold for the restrictive due date.

The dynamic programming algorithm proposed by Ventura and Weng in [244] can be easily adapted to the weighted problem. Following reasoning similar to that presented in Section 3.1.1, the recurrence relation (3.15) is developed for minimization of the weighted sum of absolute deviations with a restrictive due date. As before, $h_k^*(s)$ is the minimum cost of scheduling the first k tasks, given that the schedule starts at time s.

$$h^*_{k+1}(s) = \begin{cases} \min \{h^*_k(s + p_{k+1}) + \alpha|d - s - p_{k+1}|, \\ \quad h^*_k(s) + \beta|s + \sum_{i=1}^{k+1} p_i - d|\} & \text{if } d - s \geq 0 \\ h^*_k(s) + \beta(s + \sum_{i=1}^{k+1} p_i - d) & \text{otherwise} \end{cases} \tag{3.15}$$

The procedure is initialized with $h^*_0(s) = 0$ for all s. The minimum cost is found as follows: $\min_{0 \leq s \leq d} h^*_n(s)$.

The branch and bound algorithm and the even-odd heuristic (Algorithm 3.15) developed by Hoogeven et al. in [124] may be adapted to solve the weighted problem. However, neither the computational analysis nor the worst case performance of the heuristic are known in this case.

Bector et al. in [26] consider the WSAD problem under the assumption that due date is a decision variable. They prove that the problem is equivalent to the WSAD problem.

Parallel machines

It follows from the previous section that the WSAD problem can be tackled in a similar way to the MAD problem. Properties 3.7 – 3.9 hold. Moreover, the concept of assigning fixed weights to given positions in a schedule can be easily adopted in this case. The weight of position j calculated in the non-weighted case needs to be multiplied by α if position j occurs before the due date, and by β if the position occurs after the due date. Based on this observation, an extension of Algorithm 3.19 can be constructed to solve the weighted sum of absolute deviations problem with an unrestrictive due date. The complexity of the algorithm does not change. We present the extended algorithm below. The algorithm is formulated under the assumption that tasks are indexed according to their non-decreasing processing times.

Algorithm 3.26 (Emmons ([95])).

1. Schedule m largest tasks first, one on each machine. Set $i = m$.
2. Set $w_{ek} = \alpha/\eta_k$ and $w_{tk} = \beta/\eta_k$, $k = 1, \ldots, m$.
3. Schedule task $n - i$ in the position with the smallest weight

$$w_{xl} = \min_{k=1,\ldots,m} \{w_{ek}, w_{tk}\}.$$

4. If $x = e$ then set $w_{xl} = w_{xl} + \alpha/\eta_k$.
5. If $x = t$ then set $w_{xl} = w_{xl} + \beta/\eta_k$.

6. If $i < n$ then set $i = i + 1$ and go to step 3.

The main difference between the weighted and the non-weighted case is that the weights for particular positions differ and usually only one optimal schedule exists.

Let us consider the following example.

Example 3.27. Let us consider an instance with $n = 25$ tasks with processing times $p_1 = 1$, $p_2 = 2$, $p_3 = 4$, $p_4 = 5$, $p_5 = 7$, $p_6 = 8$, $p_7 = 10$, $p_8 = 11$, $p_9 = 12$, $p_{10} = 14$, $p_{11} = 15$, $p_{12} = 17$, $p_{13} = 18$, $p_{14} = 21$, $p_{15} = 23$, $p_{16} = 27$, $p_{17} = 32$, $p_{18} = 33$, $p_{19} = 35$, $p_{20} = 36$, $p_{21} = 38$, $p_{22} = 42$, $p_{23} = 43$, $p_{24} = 45$, and $p_{25} = 47$, and $m = 3$ machines with processing speeds $\eta_1 = 1, \eta_2 = 2, \eta_3 = 3$. Moreover, let $\alpha = 3$, $\beta = 1$, and a common due date $d = 200$. The values $\alpha||E||$ and $\beta(||T|| + 1)$ and the decisions made in consecutive iterations are presented in Table 3.7. The sequences of tasks obtained using Algorithm 3.24 are the following: $25, 12, d, 7, 17, 24, 18, 11, 3, d, 1, 6, 10, 16, 21, 23, 20, 15, 9, 5, d, 2, 4, 8, 13, 14, 19, 22$, where d marks the position of the due date. The schedule is shown in Figure 3.11.

Table 3.7. Weights of particular positions on machines in Example 3.27

		number of predecessors/successors									
Machine	set	1	2	3	4	5	6	7	8	9	10
Machine 1	E	**0**	**3**	6	9	12	15	18	21	24	27
	T	**2**	4	6	8	10	12	14	16	18	20
Machine 2	E	**0**	**1.5**	**3**	**4.5**	6	7.5	9	10.5	12	13.5
	T	**1**	**2**	**3**	**4**	**5**	6	7	8	9	10
Machine 3	E	**0**	**1**	**2**	**3**	**4**	5	6	7	8	9
	T	**0.67**	**1.33**	**2**	**2.67**	**3.33**	**4**	**4.67**	5.33	6	6.67

The problem with the restrictive due date is NP-hard, which follows immediately from the complexity of the single machine problem with restrictive due date. The complexity of the scheduling problems with the objective to minimize the weighted sum of absolute deviations is summarized in Table 3.8.

3.1.3 Symmetric weights

Let us now consider the case where the unit earliness and tardiness costs depend on the task, and $\alpha_i = \beta_i$, $i = 1, \ldots, n$. We say that such

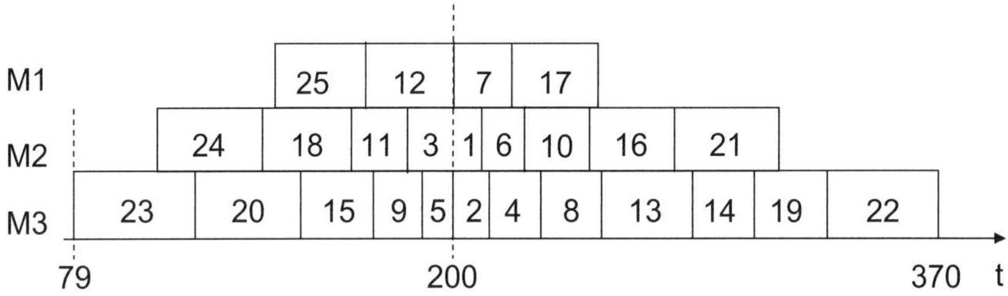

Fig. 3.11. A schedule of tasks from Example 3.27 obtained by Algorithm 3.26

Table 3.8. Complexity of WSAD problems

Problem	Complexity	Algorithms
$1\|d_j = d\|\sum(\alpha e_i + \beta t_i)$	P [205]	$O(nlogn)$ [205, 15]
$1\|d_j = d^{res}\|\sum(\alpha e_i + \beta t_i)$	NP-hard from MAD^{res}	pseudopolynomial [205, 15] enumeration [15]
$Qm\|d_j = d\|\sum(\alpha e_i + \beta t_i)$	$O(nlogn)$ [95]	
$Pm\|d_j = d^{res}\|\sum(\alpha e_i + \beta t_i)$	NP-hard from MAD^{res}	pseudopolynomial [95] enumeration [15]

weights are symmetric. For simplicity let us denote the weights by α_i, $i = 1, \ldots, n$, so the objective function is formulated as:

$$\sum_{i=1}^{n} \alpha_i |C_i - d| \tag{3.16}$$

Hall and Posner in [117] prove that this problem is NP-hard even if the due date is unrestrictive. The reduction from the even-odd partition problem is used. The even-odd partition problem is known to be NP-hard [103, 104].

Problem 3.28 (Even-odd partition problem). Consider a positive integer n and a set $X = \{x_1, \ldots, x_{2n}\}$ of positive integers, such that $x_i < x_{i+1}$ for $1 \leq i < 2n$. Does there exist a partition of X into subsets X_1 and X_2 such that $\sum_{x \in X_1} x = \sum_{x \in X_2} x$ and that for each i, $1 \leq i \leq n$, X_1 (and hence X_2) contains exactly one element of $\{x_{2i-1}, x_{2i}\}$?

Hall and Posner propose a dynamic programming algorithm and a polynomial-time approximation scheme. The proposed approximation scheme works when the maximum weight is bounded by a polynomial function of n. A fully polynomial time approximation scheme without any additional constraints imposed on the weights is presented by Kovalyov and Kubiak in [151]. De, Gosh and Wells in [80] formulate the problem with symmetric weights as a 0-1 quadratic programming problem and propose a heuristic as well as a branch and bound algorithm and dynamic programming algorithm to solve the problem. Hoogeven and van de Velde prove NP-hardness of the problem with the restrictive due date. They show two polynomially solvable cases and propose a dynamic programming algorithm for the general formulation. Alidaee and Dragan in [8] develop a polynomial time algorithm for the special case when weights are proportional to processing times. Cheng [57] presents an $O(n^{1/2}2^n)$ partial search algorithm for determining an optimal sequence and the corresponding common due date with symmetric weights. Hao et al. [119] consider the common due date determination and sequencing using a tabu search algorithm for the problem with symmetric weights. Finally, Gupta et al. in [110] propose an enumerative algorithm for the problem of minimization of the weighted earliness and tardiness costs in a two machine flow shop.

Property 3.1, known for the MAD and WSAD problems, also holds for the problem with symmetric weights (see [80, 117]). The V-shape property for the problem with symmertic weights is formulated as follows.

Property 3.12 *There exists an optimal schedule in which:*

- *the tasks finishing by the due date are scheduled in the non-increasing order of the ratio p_i/α_i, without idle time inserted;*
- *the tasks after due date are scheduled in the non-decreasing order of the ratio p_i/α_i, without idle time inserted.*

The next property (see [80, 117]) describes the optimal solution from the point of view of sums of weights in the sets of early and tardy tasks. Let us denote by $[k]$ the task scheduled in position k in sequence S.

Property 3.13 *If S is an arbitrary sequence and d is an optimal due date for S, then $d = C_{[k]}$, where k is such that*

$$\sum_{i=1}^{k-1} \alpha_{[i]} \leq \frac{1}{2}\sum_{i=1}^{n} \alpha_{[i]} \text{and} \sum_{i=1}^{k} \alpha_{[i]} \geq \frac{1}{2}\sum_{i=1}^{n} \alpha_{[i]} \tag{3.17}$$

Moreover, the following property (see [80, 117]) provides a sufficient condition for the existence of an optimal schedule in which only one task is completed by the due date.

Property 3.14 *If the first task is very long, i.e.* $p_i \geq \sum_{i=2}^{n} p_i$*, then an optimal schedule is obtained as a concatenation of sets* $E = \{1\}$ *and* $T = \{2, \ldots, n\}$ *with the task in set* E *completed exactly at the due date.*

Property 3.15 gives a sufficient condition for the existence of an optimal schedule without tardy tasks.

Property 3.15 *If* $p_k \geq \sum_{i=1}^{k-1} p_i$ *and* $\alpha_k \geq \sum_{i=k+1}^{n} \alpha_i$ *for* $k = 1, 2, \ldots, n$*, then the sets* $E = \{n, n-1, \ldots, 1\}$ *and* $T = \emptyset$ *with the last task in set* E *completed exactly at the due date define an optimal schedule.*

Some polynomially solvable special cases of the problem with unrestrictive due date are presented by De et al. in [80] and Hall and Posner in [117]. The first case occurs if the weights are equal to processing times, i.e. $\alpha_i = p_i$, $i = 1, \ldots, n$. In this case an optimal solution can be found by the following Algorithm 3.29. The algorithm assigns tasks arranged in the non-increasing order of their processing times to set E, until the sum of their processing times equals at least half of the total task processing time. We denote by P_E the total processing time of tasks assigned to set E, i.e. tasks completed by the due date.

Algorithm 3.29 (Hall and Posner [117]).

1. Set $Q = \{1, 2, \ldots, n\}$; $E = \emptyset$; $P_E = 0$.
2. Find the median value q of the set $\{p_i : i \in Q\}$.
3. Set $I = \{i \in Q : p_i = q\}$, $H := \{i \in Q : p_i > q\} \cup I_1$, such that $I_1 \subset I$ and $||H|| = \lceil ||Q||/2 \rceil$.
4. If $\sum_{i \in H} p_i + P_E \geq \lceil \sum_{i=1}^{n} p_i/2 \rceil$ then set $Q = H$, else set $E = E \cup H$, $P_E = P_E + \sum_{i \in H} p_i$, and $Q = Q \setminus H$.
5. If $Q \neq \emptyset$ then go to Step 2.
6. Set $T = \{1, 2, \ldots, n\} \setminus E$.

The computational complexity of Algorithm 3.29 is $O(n)$. Notice that the median of the set consisting of $||Q||$ elements can be found in $O(||Q||)$ time [32, 214]. Moreover, the order in which tasks are scheduled in sets E and T does not influence the value of the objective function since $p_i/\alpha_i = 1$, for $i = 1, \ldots, n$.

Another special case occurs if all the tasks have unit processing times, i.e. $p_i = 1$, $i = 1, \ldots, n$. We assume that tasks are indexed according to non-decreasing order of weights α_i, $i = 1, \ldots, n$.The optimal solution is composed of sets $E = \{n-1, n-3, \ldots, 3, 1\}$ and $T = \{2, 4, \ldots, n\}$ if n is even and $E = \{n, n-2, \ldots, 3, 1\}$ and $T = \{2, 4, \ldots, n-1\}$ if n is odd.

De et al. [80] formulate the problem of minimizing the total earliness and tardiness costs with symmetric penalties as a 0-1 programming problem and propose exact and heuristic algorithms based on this formulation. In order to present this formulation we assume that $x_i = 1$ if task i is completed by the due date and $x_i = 0$ otherwise. Then the objective function (3.16) may be transformed as follows:

$$\sum_{i=1}^{n} \alpha_i |C_i - d| = \sum_{i=1}^{n} \alpha_i \left[x_i \sum_{j>i} p_k x_k + (1 - x_i) \sum_{j>i} p_k x_k + p_j \right] \quad (3.18)$$

Exploring the above formulation, De et al. propose in [80] two branch and bound algorithms to solve the earliness/tardiness problem with symmetric weights. One of the algorithms, denoted by RBB_α, schedules tasks with smaller p_i/α_i first, and the other one, denoted by RBB_p, schedules tasks with larger p_i/α_i first.

Let us assume that tasks are indexed according to non-increasing order of the ratio p_i/α_i. A node in the enumeration tree corresponds to a partial schedule in which k tasks are already scheduled. In RBB_α, the last k tasks are scheduled to form a sequence of the following structure: $\ldots E, T \ldots$. In RBB_p, the first k tasks are scheduled to form a sequence of the following structure: $E \ldots T$. Let us denote the set of scheduled tasks by S and the set of yet unscheduled tasks by U. The cost $Z(S)$ of scheduling tasks from set S is known. The nimimum cost $Z^*(U)$ of scheduling tasks from set U may be calculated as follows. Let us denote by $p_{E'L}$ ($p_{E'U}$) a lower (upper) bound on the sum of processing times of the unscheduled tasks that are completed by the due date in an optimal schedule of $S \cup U$. The lower bound $p_{E'L}$ is calculated as the optimal solution of the following knapsack problem.

$$\begin{aligned} \text{minimize } & \sum_{i \in U} p_i x_i \\ \text{subject to } & \sum_{i \in U} \alpha_i x_i \geq \frac{1}{2} \sum_{i=1}^{n} \alpha_i - \sum_{i \in E} \alpha_i \\ & x_i \in \{0, 1\}, i = k+1, \ldots n. \end{aligned}$$

Since the knapsack problem is NP-hard, instead of solving it optimally (using existing pseudopolynomial time algorithms) a solution of the continuous relaxation of the problem is used to calculate $p_{E'L}$.

The upper bound $p_{E'U}$ is calculated as follows: $p_{E'U} = \sum_{i=k+1}^{n} p_i - p_{E'L} + \max_i\{p_i\}$.

Moreover, let us denote by $\alpha_{E'L}$ and $\alpha_{E'U}$ the lower and upper bound on the sum of weights of the unscheduled tasks that are completed by the due date in an optimal schedule of $S \cup U$. We calculate $\alpha_{E'L} = \frac{1}{2}\sum_{i=1}^{n} \alpha_i - \sum_{e \in E} \alpha_i$, and

$$\alpha_{E'U} = \begin{cases} \alpha_{E'L} + \alpha_0 & \text{if } E \neq \emptyset \\ \alpha_{E'U} = \alpha_{E'L} + \max_{i \in S \cup U}\{\alpha_i\} & \text{if } E = \emptyset \end{cases} \tag{3.19}$$

where α_0 is the last task scheduled in set E.

Three tests are performed to check if a node may be eliminated from further consideration. They are described below:

1. If at some node $\frac{1}{2}\sum_{i=1}^{n} \alpha_i < \sum_{i \in T} \alpha_i$, then the node can be eliminated from further consideration.
2. In the RBB_α algorithm consider a node at level k,
   - if task $n-k+1$ is in set E and
     $$\alpha_{n-k+1}\Big(\sum_{i \in T} p_i - \sum_{i \in E} p_i + 2p_{n-k+1}\Big) + p_{n-k+1}\Big(\sum_{i=1}^{n-k} \alpha_i - 2\alpha_{E'L}\Big) < 0$$
     then the node can be eliminated;
   - if task $n-k+1$ is in set T and
     $$\alpha_{n-k+1}\Big(\sum_{i \in T} p_i - \sum_{i \in E} p_i\Big) + p_{n-k+1}\Big(\sum_{i=1}^{n-k} \alpha_i - 2\alpha_{E'U}\Big) > 0$$
     then the node can be eliminated.
3. In the RBB_p algorithm consider a node at level k,
   - if task k is in set E and
     $$p_k\Big(\sum_{i \in T} p_i - \sum_{i \in E} p_i + 2\alpha_k\Big) + \alpha_k\Big(\sum_{i=k+1}^{n} \alpha_i - 2p_{E'L}\Big) < 0$$
     then the node can be eliminated;
   - if task k is in set T and
     $$p_k\Big(\sum_{i \in T} p_i - \sum_{i \in E} p_i\Big) + \alpha_k\Big(\sum_{i=k+1}^{n} p_i - 2p_{E'U}\Big) > 0$$
     then the node can be eliminated.

If at any mode the schedule may be completed on the basis of Properties 3.14 or 3.15 then a complete schedule is derived at this node.

If the node cannot be eliminated, two consecutive nodes are created as follows. In the RBB_α algorithm the next task (i.e. task with index $n-k$) is scheduled as the first task in set E or as the last task in set T. In the RBB_p algorithm the next task (i.e. task with index $k+1$) is scheduled as the last task in set E or as the first task in set T. A depth-first search is performed starting from the node with the smaller lower bound.

The lower and upper bounds are calculated recursively. The idea is as follows. Let us consider a partial schedule represented by a node at level k. The schedule of tasks $n-k+1, \ldots, n$, $(1, \ldots, k)$ is known in algorithm RBB_α (RBB_p). In algorithm RBB_α the lower bound on the complete schedule is calculated as follows:

$$\begin{aligned} LB_\alpha = {} & Z(S) + Z^*(U) \\ & + \sum_{i \in U} \alpha_i \sum_{i \in T} p_i + \Big(\sum_{i \in E} p_i - \sum_{i \in T} p_i \Big) \alpha_{E'L} \end{aligned}$$

and in algorithm RBB_p it is calculated as follows:

$$\begin{aligned} LB_p = {} & Z(S) + Z^*(U) \\ & + \sum_{i \in U} p_i \sum_{i \in T} \alpha_i + \Big(\sum_{i \in E} \alpha_i - \sum_{i \in T} \alpha_i \Big) p_{E'L}. \end{aligned}$$

No initial heuristic upper bound is applied, since upper bounds are calculated at each node according to the following formulas:

$$\begin{aligned} UB_\alpha = {} & Z(S) + Z^*(U) \\ & + \sum_{i \in U} \alpha_i \sum_{i \in T} p_i + \Big(\sum_{i \in E} p_i - \sum_{i \in T} p_i \Big) \sum_{i \in E^*} \alpha_i \end{aligned}$$

in algorithm RBB_p and

$$\begin{aligned} UB_p = {} & Z(S) + Z^*(U) \\ & + \sum_{i \in U} p_i \sum_{i \in T} \alpha_i + \Big(\sum_{i \in E} \alpha_i - \sum_{i \in T} \alpha_i \Big) \sum_{i \in E^*} p_i \end{aligned}$$

in algorithm RBB_p.

The algorithm RBB_α perfomes better than RBB_p in the computational tests. It solves instances with up to 40 tasks.

Pseudopolynomial time dynamic programming algorithms are also developed to solve the problem of minimizing the total earliness and tardiness costs with symmetric weights. The first two are presented by De et al. [80]. The complexity of those algorithms is $O(n \sum p_j)$ and $O(n \sum \alpha_j)$, but both proved to be very effective on the test instances reported in [80]. Another dynamic programming algorithm with time complexity $O(n \sum p_i)$ was proposed by Hall and Posner [117]. Two other dynamic programming algorithms running in $O(n \sum p_j)$ and $O(n \sum \alpha_j)$ time were proposed by Jurisch et al. [136]. The latter method is interesting because it shows a general approach to any combinatorial problem that can be transformed to the problem of finding a minimum weighted clique in a complete graph, i.e. the MinClique problem. The MinClique problem is defined as follows.

Problem 3.30 (MinClique problem). Given a complete undirected graph $G = (V, E)$ with $||V|| = n$ and weight $w_{ij} = w_{ji}$ of edge (i, j), $1 \leq i, j \leq n$ (the graph also contains edges (i, i)), find a subset V_1 of nodes such that the sum of weights of edges with both endpoints in V_1 is minimized. In other words, given matrix $[w_{ij}]$, $1 \leq j \leq i \leq n$,

$$\begin{array}{ll} \text{minimize} & \sum\limits_{1 \leq j \leq i \leq n} w_{ij} x_i x_j \\ \text{subject to} & x_i \in \{0, 1\} \end{array}$$

The weighted earliness/tardiness problem can be reduced to the MinClique problem in the following way. For a given schedule let us denote $x_i = 1$ if $C_i \leq d$, and $x_i = 0$ if $C_i > d$. The objective function (3.16) may be transformed as follows.

$$\begin{aligned} & \sum_{i=1}^{n} \alpha_i |C_i - d| \\ = & \sum_{i=1}^{n} (\alpha_i x_i \sum_{j=1}^{i-1} p_j x_j) + \sum_{i=1}^{n} \alpha_i (1 - x_i) \Big(\sum_{j=1}^{i} p_j (1 - x_j) \Big) \\ = & \sum_{1 \leq j < i \leq n} \alpha_i p_j x_i x_j + \sum_{1 \leq j \leq i \leq n} \alpha_i p_j - \sum_{1 \leq j \leq i \leq n} \alpha_i p_j x_j \end{aligned}$$

$$- \sum_{1\le j\le i\le n} \alpha_i p_j x_i + \sum_{1\le j\le i\le n} \alpha_i p_j x_i x_j$$

$$= \sum_{1\le j<i\le n} 2\alpha_i p_j x_i x_j - \sum_{i=1}^{n} \left(p_j \sum_{i=j}^{n} \alpha_i + \alpha_j \sum_{i=1}^{j-1} p_i \right) x_i x_i + \sum_{1\le j\le i\le n} \alpha_i p_j$$

$$= \sum_{1\le j<i\le n} w_{ij} x_i x_j + \sum_{1\le j\le i\le n} \alpha_i p_j$$

where

$$w_{ij} = \begin{cases} \alpha_j p_i & \text{if } i<j \\ \alpha_i p_j & \text{if } i>j \\ -\left(\alpha_i \sum_{k=1}^{i-1} p_k + p_i \sum_{k=i}^{n} \alpha_k \right) & \text{if } i=j \end{cases} \tag{3.20}$$

Concluding, the objective is to minimize the function

$$\sum_{1\le i,j\le n} w_{ij} x_i x_j. \tag{3.21}$$

A pair of row-column dynamic programming algorithms to solve the problem (3.21) can be formulated (see [136]).

In the first algorithm we assume that values x_1 to x_{k-1} have been assigned so that $\sum_{i=1}^{k-1} p_i x_i = B$, where $0 \le B \le \sum_{i=1}^{n} p_i$. Two decisions concerning the assignment of task k are possible: $x_k = 1$, meaning that task k completes by the due date, or $x_k = 0$, meaning that task k completes after the due date. Let us denote by $h_k(B, x_k)$ the minimum cost of a schedule of k tasks under the assumption that $\sum_{i=1}^{k-1} p_i x_i = B$ and the decision made is given by x_k. Moreover, let $h_k^*(B) = \min_{x_k} \{h_k(B, x_k)\}$ be the cost of optimal partial schedule of k tasks if $\sum_{i=1}^{k-1} p_i x_i = B$. Then

$$h_k(B,0) = h_{k+1}^*(B) \tag{3.22}$$

and

$$h_k(B,1) = h_{k+1}^*(B+p_k) + p_k B + w_{kk}. \tag{3.23}$$

Therefore, the recurrence relation may be formulated as follows:

$$h_k^*(B) = \min\{h_{k+1}^*(B), h_{k+1}(B+p_k) + p_k B + w_{kk}\}. \tag{3.24}$$

The initial condition is $h^*_{n+1}(B) = 0$ for all B, and the solution is obtained for $h^*_1(0)$. The complexity of the row version of the dynamic programming algorithm is $O(n\sum_{i=1}^n p_i)$.

The second algorithm in this pair of dynamic programming algorithms is formulated as follows. Let us assume that values x_{k+1} to x_n have been assigned so that $\sum_{i=k+1}^n \alpha_i x_i = A$, where $0 \le A \le \sum_{i=1}^n \alpha_i$. Two decisions concerning the assignment of task k are possible: $x_k = 1$, meaning that task k completes by the due date, or $x_k = 0$, meaning that task k completes after the due date. Let us denote by $g_k(A, x_k)$ the minimum cost of a schedule of k tasks under the assumption that $\sum_{i=k+1}^n \alpha_i x_i = A$ and the decision made is given by x_k. Moreover, let $g^*_k(A) = \min_{x_k}\{g_k(A, x_k)\}$. Then

$$g_k(A, 0) = g^*_{k-1}(A) \tag{3.25}$$

and

$$g_k(A, 1) = g^*_{k-1}(A + \alpha_k) + \alpha_k A + w_{kk}. \tag{3.26}$$

Concluding, the recurrence relation may be formulated as follows:

$$g^*_k(A) = \min\{g^*_{k-1}(A), g_{k-1}(A + \alpha_k) + \alpha_k A + w_{kk}\}. \tag{3.27}$$

The initial condition is $g^*_0(A) = 0$ for all A, and the solution is obtained for $g^*_n(0)$. The complexity of the column version of the dynamic programming algorithm is $O(n\sum_{i=1}^n \alpha_i)$.

Hall and Posner [117] develop a fully polynomial time approximation scheme for the problem with unrestrictive due date and symmetric weights when the maximum weight is bounded by a polynomial function of n. This special case is polynomially solvable which follows from the complexity of the dynamic programming procedures presented above.

The following fully polynomial-time approximation scheme was proposed by Kovalyov and Kubiak in [151]. This scheme does not require any additional constraints.

Let us assume that the variables x_i, $i = 1, \ldots, n$, are defined as in the 0-1 mathematical programming formulation. The set of all j-element partial schedules is denoted by $X_j = \{(x_1, \ldots, x_n) : x_i = 0, i = j+1, \ldots, n\}$. The approximation scheme proposed by Kovalyov and Kubiak uses the algorithm *Partition*(A, G, δ), where $A \subseteq X_j$, G is a nonnegative integer function on X_j, and $0 \le \delta \le 1$. The algorithm finds a partition of A into disjoint subsets $A_1^G, \ldots, A_k^G$, such that $|G(\mathbf{x}) - G(\mathbf{x}')| \le \delta \min\{G(\mathbf{x}), G(\mathbf{x}')\}$ for any $\mathbf{x}, \mathbf{x}'$ from the same subset A_i^G, $i = 1, \ldots, k$. The algorithm is presented below.

Algorithm 3.31 (Partition(A, G, δ), [151]).

1. Arrange vectors $\mathbf{x} \in A$ in order $\mathbf{x}^1, \ldots, \mathbf{x}^{||A||}$, where $0 \leq G(\mathbf{x}^1) \leq \ldots \leq G(\mathbf{x}^{||A||})$.
2. Set $k = 1$ and $i_0 = 1$.
3. Assign vectors $\mathbf{x}^{i_{(k-1)}+1}, \mathbf{x}^{i_{(k-1)}+2}, \ldots, \mathbf{x}^{i_k}$ to set A_k^G until for some i_k the following inequalities hold: $G(\mathbf{x}^{i_k}) \leq (1+\delta)G(\mathbf{x}^{i_{(k-1)}+1})$ and $G(\mathbf{x}^{i_k+1}) > (1+\delta)G(\mathbf{x}^{i_{(k-1)}+1})$. If such an i_k does not exist, then take $A_1^G = A$ and stop.
4. If $i_k < ||A||$, then set $k = k+1$ and go to step 3.

Let us denote by $\mathbf{e_j}$ an n-element vector with all entries equal to zero, except for $e_j = 1$. The algorithm builds a sequence of sets $Y_0, \ldots, Y_n$, where $Y_j \subseteq X_j$. Set Y_0 contains a single vector $(0, \ldots, 0)$. From each vector $\mathbf{x}$ in set Y_{j-1} two vectors are created by adding 0 or 1 in position j of $\mathbf{x}$. Then the procedure *Partition* is used to partition Y_j into disjoint subsets in such a way that for any two solutions $\mathbf{x}$, $\mathbf{x}'$ in the same subset the solutions are close with respect to the values of $W_j(\mathbf{x})$, $P_j^E(\mathbf{x})$ and $P_j^T(\mathbf{x})$, respectively. Solutions $\mathbf{x}$ and $\mathbf{x}'$ are close with respect to $W_j(\mathbf{x})$ if

$$|W_j(\mathbf{x}) - W_j(\mathbf{x}')| \leq \varepsilon \min\{W_j(\mathbf{x}, W_j(\mathbf{x}')\}.$$

Solutions $\mathbf{x}$ and $\mathbf{x}'$ are close with respect to $P_j^E(\mathbf{x})$ if

$$|P_j^E(\mathbf{x}) - P_j^E(\mathbf{x}')| \leq (\varepsilon/2n) \min\{P_j^E(\mathbf{x}), P_j^E(\mathbf{x}')\}.$$

Solutions $\mathbf{x}$ and $\mathbf{x}'$ are close with respect to $P_j^T(\mathbf{x})$ if

$$|P_j^T(\mathbf{x}) - P_j^T(\mathbf{x}')| \leq (\varepsilon/2n) \min\{P_j^T(\mathbf{x}), P_j^T(\mathbf{x}')\}.$$

From each subset two solutions are selected to be included in set Y_j, the one with maximum $W_j(\mathbf{x})$ and the one with minimum $W_j(\mathbf{x})$. All other solutions are discarded. Kovalyov and Kubiak show that in this way a solution with value not exceeding $(1+\varepsilon)W^*$ is obtained in the last iteration.

Algorithm 3.32 (Algorithm H_ε, [151]).

1. Index the tasks so that $p_1/\alpha_1 \leq \ldots \leq p_n/\alpha_n$. Set $Y_0 = \{0, \ldots, 0\}$, and $j = 1$.
2. For each $\mathbf{x} \in Y_{j-1}$ set $Y_j' = Y_{j-1} \cup \{\mathbf{x} + \mathbf{e_j} : \mathbf{x} \in Y_{j-1}\}$.

3. For each $\mathbf{x} \in Y_j^{'}$ calculate
$P_j^E(\mathbf{x}) = P_{j-1}^E(\mathbf{x}) + p_j x_j$,
$P_j^T(\mathbf{x}) = \sum_{i=1}^{j} p_i - P_j^E(\mathbf{x})$,
$W_j(\mathbf{x}) = \begin{cases} W_{j-1}(\mathbf{x}) + \alpha_j P_{j-1}^E(\mathbf{x}) \text{ if } x_j = 1 \\ W_{j-1}(\mathbf{x}) + \alpha_j P_j^T(\mathbf{x}) \quad \text{if } x_j = 0 \end{cases}$
where $W_0(\mathbf{x}) = 0$.
4. If $j = n$ then set $Y_n = Y_n^{'}$ and go to step 8.
5. If $j < n$ then
perform algorithm $Partition(Y_j^{'}, W_j, \varepsilon)$ to obtain $Y_1^W \dots Y_{k_W}^W$;
perform algorithm $Partition(Y_j^{'}, P_j^E, \varepsilon/2n)$ to obtain $Y_1^E \dots Y_{k_E}^E$;
perform algorithm $Partition(Y_j^{'}, P_j^T, \varepsilon/2n)$ to obtain $Y_1^T \dots Y_{k_T}^T$.
6. For all triples (l_1, l_2, l_3) such that $Y_{l_1}^W \cap Y_{l_2}^E \cap Y_{l_3}^T \neq \emptyset$, $1 \leq l_1 \leq k_W$, $1 \leq l_2 \leq k_E$ and $1 \leq l_3 \leq k_T$, set $Z_{(l_1,l_2,l_3),j} = Y_{l_1}^W \cap Y_{l_2}^E \cap Y_{l_3}^T$. In each subset $Z_{q,j}$ choose vectors $\mathbf{x}^{\mathbf{q},\mathbf{min}}$ and $\mathbf{x}^{\mathbf{q},\mathbf{max}}$ such that

$$W_j(\mathbf{x}^{\mathbf{q},\mathbf{min}}) = \min\{W_j(\mathbf{x}) : \mathbf{x} \in Z_{q,j}\},$$

$$W_j(\mathbf{x}^{\mathbf{q},\mathbf{max}}) = \max\{W_j(\mathbf{x}) : \mathbf{x} \in Z_{q,j}\}.$$

7. Set $Y_j = \{\mathbf{x}^{\mathbf{q},\mathbf{min}}, \mathbf{x}^{\mathbf{q},\mathbf{max}} : q$ is a triplet (l_1, l_2, l_3) such that $Y_{l_1}^W \cap Y_{l_2}^E \cap Y_{l_3}^T \neq \emptyset\}, 1 \leq l_1 \leq k_W, 1 \leq l_2 \leq k_E$ and $1 \leq l_3 \leq k_T\}$. Set $j = j + 1$ and go to step 2.
8. Select vector $\mathbf{x_0} \in Y_n$, such that $W_n(\mathbf{x_0}) = \min\{W_n(\mathbf{x}) : \mathbf{x} \in Y_n\}$

The complexity of the polynomial approximation scheme is $O(n^2 log^3 (\max_j\{n, 1/\varepsilon, p_i, w_j\})\ /\varepsilon^2)$.

Van den Akker et al. in [239] transform the earliness/tardiness problem into the *set covering problem* and present a combined column generation and Lagrangian relaxation algorithm that finds optimal solutions to instances with up to 125 tasks.

The following heuristic algorithm to solve problem (3.18) is proposed by De at al. in [80]. Let us denote by Δ_i the coefficient of variable x_i in the objective function (3.18). The coefficient is defined as follows:

$$\Delta_i = 2p_i \sum_{k<i} \alpha_k x_k + 2\alpha_i \sum_{k>i} p_k x_k - p_i \sum_{k<i} \alpha_k - \alpha_i p_i - \alpha_i \sum_{k>i} p_k. \quad (3.28)$$

The computational complexity of the heuristic algorithm, given below, is $O(n^2)$.

Algorithm 3.33 (QP-H, [80]).

1. Set $E = \{1, 2 \ldots, n\}$, and $T = \emptyset$, i.e. $x_i = 1$, $i = 1, \ldots, n$.
2. Compute Δ_i, for $i = 1, \ldots, n$.
3. If $\Delta_i \le 0$ for all $i \in E$ and $\Delta_i \ge 0$ for all $i \in T$ then stop, else find j such that

$$\Delta_j = \max_i \{|\Delta_i| : i \in E \,\text{and}\, \Delta_i > 0 \,\text{or}\, i \in T \,\text{and}\, \Delta_i < 0\}$$

4. If $j \in E$ then set $x_j = 0$, else set $x_j = 1$ and go to Step 2.

De et al. in [80] report the results of a computational experiment showing that Algorithm 3.33 finds solutions close to optimal (less that 1% deviation) in a very short time.

A tabu search algorithm to solve the problem with symmetric weights was proposed by Hao et al. in [119]. The computational experiments show that the tabu search algorithm is even faster than Algorithm 3.33 and yields slightly better results. The advantage of the tabu search algorithm over Algorithm 3.33 grows with the instance size.

Hoogeven and van de Velde prove in [123] that the problem of minimization of the total earliness and tardiness costs with symmetric penalties and a restrictive due date is NP-hard. They show two special cases which are polynomially solvable and propose a dynamic programming algorithm with complexity $O(n^2 d)$, which solves the problem in the general case.

Property 3.12 holds for optimal schedules of the problem with restrictive due date. Moreover, the following property is proved in [123].

Property 3.16 *In every optimal schedule, either the first task starts at time 0, or the due date d coincides with the start time or completion time of the task with the smallest p_i/α_i ratio.*

Let us consider the dynamic programming algorithm presented in [123]. Two cases distinguished by Property 3.16 are considered. In the first case, the task with the largest value of α_i/p_i is scheduled so that its completion or start time coincides with the due date. We index the tasks according to the non-decreasing order of the ratios p_i/α_i, $i = 1, \ldots, n$. We denote by $F_j(t)$ the optimal cost of a schedule with the interval $[d - t, d + \sum_{i=1}^{j} p_i - t]$ occupied by the first j tasks. The recurrence relation of the dynamic programming algorithm is formulated as follows:

$$F_j(t) = \min\Big\{F_{j-1}(t-p_j) + \alpha_j(t-p_j),$$

$$F_{j-1}(t) + \alpha_j\Big(\sum_{j=1}^{j} p_i - t\Big)\Big\} \text{ for } 0 \leq t \leq d.$$

The initial conditions are:

$$F_j(t) = \begin{cases} 0 & \text{for } t = 0, j = 0, \\ \infty & \text{otherwise.} \end{cases}$$

In the second case, all tasks are completed in the interval $[0, \sum_{i=1}^{n} p_i]$. Let us now index the tasks according to the non-increasing order of the ratios p_i/α_j. The first task either starts at time zero or finishes at time $\sum_{i=1}^{n} p_i$. We denote by k the task which starts before and completes after the due date and by $G_j^k(t)$ the optimal cost of scheduling the first j tasks under the condition that the scheduled tasks occupy the intervals $[0, t]$ and $[\sum_{i=j+1}^{n} p_j + t, \sum_{i=1}^{n} p_i]$.

The recurrence relation is formulated as follows:

$$G_j^k(t) = \begin{cases} G_{j-1}^k(t) & \text{if } j = k \\ G_{j-1}^k(t) + \alpha_j\Big(\sum_{i=j}^{n} p_i + t - d\Big) & \text{if } d - j \leq p_i \leq t \leq d \\ G_{j-1}^k(t-p_j) & \text{if } \sum_{i=j+1}^{n} p_i < d - t \\ \min\Big\{G_{j-1}^k(t) + \alpha_j\Big(\sum_{i=j}^{n} p_i + t - d\Big), & \\ G_{j-1}^k(t-p_j) + \alpha_j(d-t)\Big\} & \text{otherwise} \end{cases}$$

The initial conditions are:

$$G_j^k(t) = \begin{cases} 0 & \text{for } t = 0, j = 0 \\ \infty & \text{otherwise.} \end{cases}$$

Taking into consideration that task k is scheduled in the interval $[t, t + p_k]$, the final cost is calculated as

$$G_n^h(t) = \begin{cases} G_n^k(t) + \alpha_k(t + p_k - d) & \text{if } d - p_k \leq t \leq d \\ \infty & \text{otherwise.} \end{cases}$$

Finally, the minimum cost is found as follows:

$$\min\{\min_{1\le k\le n}\{\min_{d-p_k\le t\le d} G_n^k(t)\}, \min_{0\le t\le d} F_n(t)\}.$$

There are two special cases of the problem with symmetric weights, for which optimal solutions can be found in polynomial time. The first one occurs if all tasks have identical processing times, i.e. $p_i = p$, $i = 1, \dots, n$. We assume that tasks are indexed according to the non-increasing order of their weights α_i. Since the processing times are equal for all tasks, we can assign a weight to each position in the schedule. In an optimal schedule, tasks with the largest α_i are assigned to positions with the smallest weights. The following algorithm finds an optimal schedule in $O(nlogn)$ time.

Algorithm 3.34 (Hoogeveen and van de Velde [123]).

1. If $d \ge p\lceil\frac{1}{2}n\rceil$ then set $E = \{1, 3, \dots, 2\lceil\frac{1}{2}n\rceil - 1\}$ and $T = \{2, 4, \dots 2\lfloor\frac{1}{2}n\rfloor\}$. Order tasks in E and T according to Property 3.12 and start the schedule (E, T) at time $d - p\lceil\frac{1}{2}n\rceil$.
2. If $d < \lceil\frac{1}{2}n\rceil$ then choose the better one of the two following solutions: start the schedule at time 0, or complete the last task in E at time d.

The second special case occurs when the ratios of weights to processing times are equal for all tasks, i.e. $p_i/\alpha_i = p_j/\alpha_j$, $i, j = 1, \dots, n$. In this case there exists an optimal schedule in which tasks are completed in the order of non-increasing processing times. This property holds for a restrictive as well as unrestrictive due date. An algorithm of complexity $O(nlogn)$ to solve this problem is proposed by Alidaee and Dragan in [8].

Algorithm 3.35 (Alidaee and Dragan [8]).

1. Order the tasks according to non-increasing processing times to obtain sequence S.
2. Construct schedule S_0 by scheduling tasks from sequence S without processor idle time between tasks and starting the first task at time 0.
3. For all i such that $C_i < d$ in schedule S_0, find a schedule S_i, by scheduling tasks from sequence S without processor idle time between tasks, so that task i completes at time d.
4. Choose from the schedules S_i, $i = 0, \dots, k$, where k is the number of tasks completed by d in schedule S_0, the schedule with minimum cost.

The results discussed so far in this section concern problems of scheduling tasks on a single machine.

Very few studies deal with multiple machine systems. Achuthan et al. in [2] consider a two machine flow shop with minimization of the maximum earliness and tardiness costs. They define earliness as a monotonous function of the difference of the desired and the actual starting time of the operation on the first machine.

The problem with symmetric weights is also considered with the objective to minimize the maximum weighted absolute lateness, i.e.

$$\max_{1 \le i \le n} \{\alpha_i |C_i - d|\}. \tag{3.29}$$

For a given due date the problem is NP-hard (see [174]). However, in the case of unit weights Lakshminarayan et al. in [166] provide an $O(nlogn)$ algorithm to solve the problem.

3.1.4 Total Weighted Earliness and Tardiness

In this section we consider the problem with a common due date and task dependent linear earliness and tardiness costs, i.e. with the objective to minimize function (3.1). This problem is often referred to as the total weighted earliness and tardiness (TWET) problem. Since the problem with unrestrictive due date and symmetric weights is NP-hard [117], the TWET problem is also NP-hard.

In addition to Properties 3.5 and 3.7 , the following properties hold for the optimal schedules for the TWET problem.

Property 3.17 *In an optimal schedule for the TWET problem the tasks completed by the due date are ordered according to the non-increasing ratio* p_i/α_i*, and tasks completed after the due date are ordered according to the non-decreasing ratio* p_i/β_i*.*

Property 3.18 *If* d *is unrestrictive, then in an optimal schedule the* k*-th task completes at time* d*, where* k *is the smallest integer satisfying the inequality:*

$$\sum_{j=1}^{k} (\alpha_j + \beta_j) \ge \sum_{j=1}^{n} \beta_j. \tag{3.30}$$

Dileepan et al. in [89] propose a branching scheme for an enumeration procedure in the case of unrestrictive due date. On this basis they develop a heuristic procedure to solve the TWET problem. It follows from Property 3.17 that if we know the assignment of tasks to sets

E and T then it is easy to construct an optimal schedule. Thus the branching procedure enumerates all possible assignments of tasks to sets E and T as follows. In the root of the branching tree all tasks are assigned to set E, and set T is empty. Tasks in set E are ordered according to the non-increasing ratio p_i/α_i. New nodes are created from a given one by removing each task from set E and placing it as the first task in set T. If the new assignment of tasks violates Property 3.18 or if the order of tasks in set T is not consistent with Property 3.17, then the node is discarded. The branching is continued until no new nodes are created. A partial schedule at each node is created by scheduling a concatenation of E and T in such a way that the last task in set E is completed at time d. The value of the objective function is calculated at every node and the best solution is chosen.

Let us illustrate this scheme with an example.

Example 3.36. Consider a set of $n = 5$ tasks with processing times and unit earliness and tardiness penalties given in Table 3.9.

Table 3.9. Data for Example 3.36

Task	1	2	3	4	5
p_i	24	6	12	11	5
α_i	4	2	7	8	4
β_i	2	9	7	8	5
p_i/α	3.00	6.00	1.71	1.38	1.25
p_i/β	12.00	0.67	1.71	1.38	1.00

The branching tree is presented in Figure 3.12. Only the tasks in set T are shown at each node. The number outside of each node is the schedule cost.

In the optimal schedule set $E = \{2, 3, 4\}$, $T = \{5, 1\}$ and the schedule cost equals 206.

The heuristic algorithm proposed in [89] is based on the branching tree. If moving a task from set E to set T violates Property 3.18, then such move is not considered. At most one descendant of each node is expanded. It is the node corresponding to the task for which the cost K of removing the task from set E and adding it to set T, in the position preserving the V-shape understood as in Property 3.17, is the biggest of all neighbors for which this cost is positive. Let us assume that task i removed from set E is placed in set T directly after task j. If tasks are

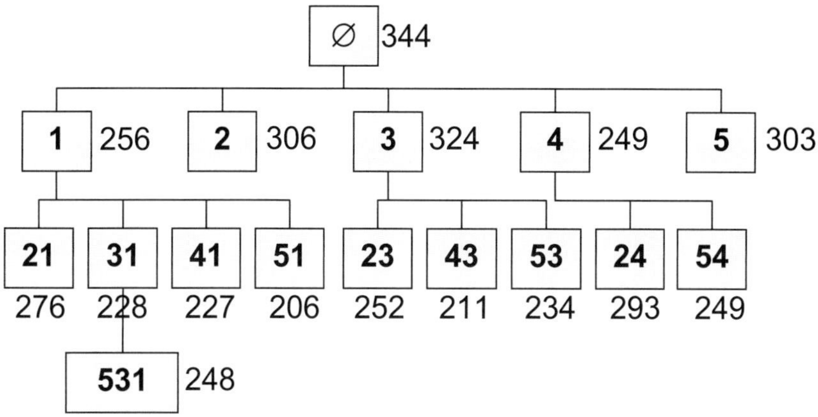

Fig. 3.12. A branching tree for Example 3.36

indexed according to the order in the schedule represented by the parent node, with l tasks in set E, then the cost K is calculated according to formula (3.31).

$$K = p_i \left[\sum_{k=1}^{i-1} \alpha_k - \sum_{k=j+1}^{n} \beta_k \right] + \alpha_i \sum_{k=i+1}^{l} p_k - \beta_i \left(\sum_{k=1}^{j} p_k + p_i \right) \quad (3.31)$$

The heuristic algorithm is presented below.

Algorithm 3.37 (Dileepan [89]).

1. Set $E = \{1, \ldots, n\}$ and $T = \emptyset$.
2. Arrange the tasks in set E according to the non-increasing ratio p_i/α_i.
3. If there is no task that can be removed from set E and inserted into set T without violating inequality (3.30) then stop.
4. For each task from set E calculate the value K of moving the task to set T, according to formula (3.31). If all the values K are non-positive then stop.
5. Select the task with the biggest positive value K and insert it to set T, preserving the non-decreasing ratio p_i/β_i and go to step 3.

The computational experiment reported in [89] shows that instances with 15 tasks can be solved optimally with the branching scheme. The heuristic solutions differ by less that 2% from the optimum and the computational time of the heuristic is 90% smaller than of the enumerative algorithm.

Another heuristic algorithms for the TWET problem is proposed by De et al. in [84]. It is a greedy randomized adaptive search procedure. The algorithm consists in two phases. In the first phase an initial solution is constructed randomly and in the second phase this solution is improved using the deepest descent neighborhood search.

Biskup and Feldman [29] and Biskup [27] develop a problem generator to obtain benchmark instances for the TWET problem, taking into account more or less restrictive due date. They further compare two heuristic algorithms on the basis of a computational experiment. One of the heuristics is a slightly modified Algorithm 3.37 adjusted to the case of any restrictive due date. The procedure starts with all tasks in set T and moves the tasks to set E. A move is infeasible if after adding the consecutive task the total processing time of tasks in set E exceeds d. The idea that underlines the second heuristic is to schedule tasks with relatively large tardiness costs early. The tasks are indexed according to decreasing ratios α_i/β_i and the ties are solved by scheduling the task with larger β_i first. The algorithm is presented below.

Algorithm 3.38 (Biskup and Jahnke [30]).

1. Set $k = 1$, $E = \emptyset$, $T = \{1, 2, \ldots, n\}$.
2. If $||E|| \geq \lceil n/2 \rceil$ then go to step 5.
3. Consider the k-th task in set T. If $d - \sum_{i \in E} p_i \geq p_k$ then $E = E \cup \{k\}$.
4. If $k \geq n$ then set $T = T \setminus E$ and go to step 7, else set $k = k + 1$ and go to step 2.
5. Set $T = T \setminus E$ keeping the order of tasks in T. Let l be the first task in set T. If the value of the objective function for schedule $(E \cup \{l\}, T \setminus \{l\})$ is larger or equal to the one for schedule (E, T) then go to step 7, else set $E = E \cup \{l\}$, $T = T \setminus \{l\}$.
6. If $||T|| > 0$ then go to step 5.
7. Arrange the tasks in set E in the non-increasing ratio p_i/α_i and in set T in the non-decreasing ratio p_i/β_i. Calculate the value of the objective function and stop.

The computational experiments reported in [29] show that Algorithm 3.38 performs better than modified Algorithm 3.37, especially for larger instances of the problem.

James [128] proposes to use tabu search for the total weighted earliness and tardiness problem with restrictive and unrestrictive due date. It is worth noticing that James restricts the search only to sequences starting at time zero. Thus an optimal sequence may be excluded from the search process.

Feldman and Biskup in [96] compare several metaheuristic approaches, including evolutionary strategy (with and without a destabilization phase), threshold acceptance algorithm (with and without a back step), and a simulated annealing algorithm used to solve the TWET problem. They present the results of computational experiments performed on the set of benchmark instances proposed in [29]. Hino et al. in [121] develop a dedicated heuristic and apply a simulated annealing, a tabu search, and a genetic algorithm, and two hybrid metaheuristics combining a genetic algorithm and a tabu search algorithm to solve the TWET problem. They compare the proposed algorithms in a computational experiment performed on the set of benchmark instances given in [29]. The best results are reported for one of the hybrid metaheuristics, which applies a genetic algorithm and then tabu search with the best solution found by GA as initial solution. For large instances, this heuristic outperforms the threshold acceptance algorithm with a back step, which was the best of the algorithms analyzed in [96].

Lee et al. [170] present a pseudopolynomial time algorithm for a special case of the TWET problem with agreeable ratios, which means that $p_i/\alpha_i < p_j/\alpha_j$ implies $p_i/\beta_i < p_j/\beta_j$ for all i and j. Lee et al. consider schedule cost including not only the earliness and tardiness costs, but also the weighted cost of task completion times. Szwarc [234] proposes several improvements to this procedure.

Balakrishnan et al. in [18] consider the problem of scheduling tasks with sequence-dependent setup times on uniform parallel machines so as to minimize the total earliness and tardiness costs. They formulate the problem as a mixed-integer program (3.32) - (3.40). To present the program let us denote by y_{im}, $i = 1, \ldots, n$, $m = 1, \ldots, M$, a decision variable equal to 1 if task i is processed on machine j and 0 otherwise, by x_{ij} a decision variable equal to 1 if task i precedes task j on the same machine and 0 otherwise, by s_{ijm} the setup time that occurs if task i precedes task j on machine m, where $i = 0$ corresponds to machine idle time, and by L a large number. Since machines are uniform, it is assumed that processing times as well as setup times on particular machines are proportional, i.e. $p_{im} = \eta_m p_{i1}$ and $s_{ijm} = \eta_m s_{ij1}$. Finally, r_i denotes the ready time of task i, $i = 1, \ldots, n$. The problem is then formulated as follows:

minimize

$$\sum_{i=1}^{n} (\alpha_i e_i + \beta_i t_i) \tag{3.32}$$

subject to

$$\sum_{m=1}^{M} y_{im} = 1, i = 1, \ldots, n \tag{3.33}$$

$$y_{im} + \sum_{m' \neq m} y_j m' + x_{ij} \leq 2, i = 1, \ldots, n-1; m = 1, \ldots, M \tag{3.34}$$

$$C_i + e_i - t_i = d_i, i = 1, \ldots, n \tag{3.35}$$

$$C_j - C_i + L(3 + x_{ij} - y_{im} - y_{jm}) \geq p_i m + s_{ijm}, \tag{3.36}$$

$$i = 1, \ldots, n-1; j = i+1, \ldots, n; m = 1, \ldots, M$$

$$C_i - C_j + L(2 + x_{ij} - y_{im} - y_{jm}) \geq p_{im} + s_{jim}, \tag{3.37}$$

$$i = 1, \ldots, n-1; j = i+1, \ldots, n; m = 1, \ldots, M$$

$$C_i \geq r_i + p_{im} y_{im}, i = 1, \ldots, n; m = 1, \ldots, M \tag{3.38}$$

$$C_i \geq s_{0im} y_{im} + p_{im} y_{im}, i = 1, \ldots, n; m = 1, \ldots, M \tag{3.39}$$

$$e_i, t_i, C_i \geq 0, i = 1, \ldots, n \tag{3.40}$$

Constraint (3.33) assigns each task to exactly one machine and constraint (3.34) assures that task i is considered as a predecessor of task j only if both tasks are assigned to the same machine. Constraint (3.35) defines the earliness and tardiness of task i. Disjunctive constraints (3.36) and (3.37) establish completion times of tasks i and j assigned to the same machine. Constraints (3.38) and (3.39) ensure that task ready times are respected. Finally, non-negativity of variables is guaranteed by constraint (3.40).

Balakrishnan et al. in [18] report results of a computational experiment where the mixed-integer program is solved using the LINDO package. Obviously, this approach can be applied only to small instances. For large problem instances Balakrishnan et al. propose an algorithm based on Bender's decomposition (see [259]). Unfortunately, using this algorithm, it is only possible to solve problems with up to 12 tasks and 3 machines on a 333MHz Pentium based PC.

An enumerative algorithm for the two-machine flow shop with restrictive as well as unrestrictive due date is proposed by Gupta et al. in [110]. The objective function considered in [110] is the weighted sum of arbitrary functions of task earliness and tardiness. The proposed enumerative algorithm solves instances with up to 20 tasks.

3.1.5 Controllable due date

Earlier in this chapter we assumed that the due date is given in the problem formulation. However, in many situations the due date may be negotiated with the customer. The scheduling problem where the due date is assumed to be a decision variable is called the due date assignment problem. In early papers, the computer simulation techniques were applied to determine the due dates (see [72, 93, 250, 249]). Classification of due date assignment problems is presented in the Table 2.2. In the case where all tasks have a common due date, the problem is abbreviated as CON. Gordon et al. in [108] present an extensive survey of results concerning the CON problem. It is easy to notice that an optimal solution to the CON problem may be obtained from an optimal solution of the problem with unrestrictive due date by setting the due date at $d = \sum_{i \in E} p_i$.

In this section we consider problems where the goal of minimizing the due date is explicitly expressed in the objective function formulation. Panwalkar et al. in [205] were the first to examine the earliness/tardiness problem together with the due date assignment. They consider the problem of minimizing the following objective function

$$\sum_{i=1}^{n} (\alpha e_i + \beta t_i + \gamma d). \tag{3.41}$$

The following properties of optimal schedules are presented in [205]. First, the optimal due date is not larger than the total processing time of all tasks.

Property 3.19 *The optimal due date fulfills the following inequality*

$$d \le \sum_{i=1}^{n} p_i.$$

Moreover, for large γ the optimal due date is equal to 0.

Property 3.20 *If $\gamma > \beta$ then in an optimal schedule the tasks are ordered according to non-decreasing processing times, and the optimal due date is equal to 0.*

Otherwise, in each sequence of tasks one task is completed at d.

Property 3.21 *For any given sequence of tasks, there exists an optimal due date equal to completion time C_k of task k, where $k = \lceil n(\beta-\gamma)/(\alpha+\beta)\rceil$.*

Let us assume that k is the task completed exactly at the due date. Then function (3.41) can be rewritten as

$$\begin{aligned}
&\sum_{i=1}^{n} \alpha e_i + \beta t_i + \gamma d \\
&= \sum_{i=1}^{k} (n\gamma + (i-1)\alpha)p_i + \sum_{i=k+1}^{n} \beta(n+1-i)p_i \\
&= \sum_{i=1}^{n} w_i p_i,
\end{aligned} \tag{3.42}$$

where

$$w_i = \begin{cases} n\gamma + (i-1)\alpha & \text{for } 1 \le i \le k, \\ \beta(n+1-i) & \text{for } k < i < n. \end{cases} \tag{3.43}$$

Thus, the cost of scheduling a task depends on its position in the sequence. In an optimal sequence the task with the largest processing time is scheduled in the position with the smallest weight, the task with the second largest processing time in the position with the second smallest weight etc. This property follows from a well known result in linear algebra [120]. On the basis of this property an $O(nlogn)$ algorithm was proposed in [205] to solve problem (3.41).

Algorithm 3.39 (Panwalkar et al. [205]).

1. Set $k = n(\beta - \gamma)/(\alpha + \beta)$.
2. If $k \le 0$ then set $d = 0$, order tasks according to non-decreasing processing times and stop, else set $k = \lceil k \rceil$.
3. Set $w_i = n\gamma + (i-1)\alpha$ for $i = 1, \ldots, k$.
4. Set $w_i = \beta(n+1-i)$ for $i = k+1, \ldots, n$.
5. Put the weights w_i in non-increasing order. Let $r(i)$ be the position of w_i in this order.

6. Obtain the optimal sequence by putting task $r(i)$ in position i.
7. Set $d = \sum_{i=1}^{k} p_{r(i)}$.

Let us illustrate the algorithm with an example.

Example 3.40. Let us consider $n = 7$ tasks with processing times $p_1 = 2$, $p_2 = 3$, $p_3 = 5$, $p_4 = 6$, $p_5 = 8$, $p_6 = 11$, and $p_7 = 15$. The cost coefficients are the following: $\alpha = 10$, $\beta = 17$, and $\gamma = 5$.

First, from Property 3.21, we obtain $k = 4$. The position weights are shown in Table 3.10.

Table 3.10. Position weights for Example 3.40

Position	1	2	3	4	5	6	7
w_i	35	45	55	65	51	34	17
$r(i)$	5	4	2	1	3	6	7

It follows from the obtained position ranking that the optimal sequence is (5, 4, 2, 1, 3, 6, 7), $d = 19$, and the total cost equals 1159.

The above reasoning can be also applied to the problem with objective function (3.44) which includes the cost of the flow time.

$$\sum_{i=1}^{n} (\alpha e_i + \beta t_i + \gamma d + \delta C_i). \tag{3.44}$$

Panwalkar et al. show in [205] that properties 3.19 – 3.21 hold in this case. Thus, the number of tasks completed by the due date is calculated as in Property 3.21. The optimal sequence of tasks is constructed as in Algorithm 3.39, taking into account that each position weight increases by $\delta(n + 1 - i)$, $i = 1, \ldots, n$. The position weights are now calculated as follows:

$$w_i = \begin{cases} n\gamma + (i-1)\alpha + (n+1-i)\delta & \text{for } 1 \leq i \leq k, \\ \beta(n+1-i) + (n+1-i)\delta & \text{for } k < i < n. \end{cases} \tag{3.45}$$

Another extension of the due date assignment problem (3.41) proposed by Panwalkar et al. in [205] is to minimize the following objective function:

$$\sum_{i=1}^{n} (\alpha e_i + \beta t_i + \gamma \max\{0, d - A\}). \tag{3.46}$$

where A is an acceptable lead time, meaning that the due date penalty is applicable only if d exceeds A. If $A > 0$ then this problem is NP-hard.

Cheng [54, 56] shows that the optimal due date can be set at $d = C_k + \delta p_{k+1}$, where $k = n(\beta - \gamma)/(\alpha + \beta)$ and $0 \leq \delta \leq 1$. Thus, there exist an infinite number of optimal due dates.

In [55] Cheng considers the problem with the objective function (3.41) in a system of identical parallel machines. He shows that if a sequence of tasks on each machine is given, and the tasks are indexed according to non-decreasing completion times, then an optimal due date is set as $d = C_k$, where k satisfies $k = \lceil n(\beta - \gamma)/(\alpha + \beta) \rceil$. Cheng and Chen in [59] and De et al. in [85] show that this problem is NP-hard even for two machines. Moreover, De et al. prove in [85] that it is strongly NP-hard for arbitrary m. To solve this problem Cheng proposes in [55] a heuristic (Algorithm 3.41) which preserves the V-shape property of the optimal schedule.

Algorithm 3.41 (Cheng [55]).

1. Index tasks according to non-decreasing order of processing times.
2. Set $l = \lceil n/m \rceil$ and $k = \lceil l(\beta - \gamma)/(\alpha + \beta) \rceil$.
3. Set $\lambda_{ij} = l\gamma + (i-1)\alpha$ for $1 \leq i \leq k$, $1 \leq j \leq m$ and $\lambda_{ij} = (l+1-i)\beta$ for $k+1 \leq i \leq l$, $1 \leq j \leq m$.
4. Put values λ_{ij} in non-increasing order.
5. Assign the first task to the position corresponding to the first value λ_{ij}, the second task to the position corresponding to the second value λ_{ij}, etc.
6. Find an optimal schedule of tasks on each machine separately and re-index the tasks according to non-decreasing completion times.
7. Calculate $k^* = \lceil n(\beta - \gamma)/(\alpha + \beta) \rceil$ in order to find the optimal due date given by $d^* = C_{k^*}$. Calculate the total cost of this schedule as follows: $\sum_{j=1}^{m} \sum_{i=1}^{n} (\alpha_i e_{ij} + \beta_i t_{ij}) + \gamma d$.

Diamond and Cheng propose in [88] a modification of Algorithm 3.41 in which $l = n/m$. They assume that the schedule starts at time zero on at least one machine. In consequence, on each machine one task is completed exactly at the due date. Diamond and Cheng show that the worst case performance of the heuristic is less than $2(m-1)\beta^2/n\gamma^2$.

Let us illustrate Algorithm 3.41 with a numerical example.

Example 3.42. Let us consider the set of $n = 8$ tasks with processing times $p_1 = 2$, $p_2 = 3$, $p_3 = 4$, $p_4 = 6$, $p_5 = 7$, $p_6 = 8$, $p_7 = 9$, and $p_8 = 11$ to be scheduled on $m = 3$ machines. The cost coefficients are $\alpha = 4$, $\beta = 5$, and $\gamma = 2$. The tasks are already indexed as required, so we pass to step 2 of the algorithm and calculate $l = \lceil 8/3 \rceil = 3$ and $k = 3 * (5-2)/(4+5) = 1$. Since $k = 1$, in step 3 we obtain $\lambda_{11} = \lambda_{12} = \lambda_{13} = 6$, $\lambda_{21} = \lambda_{22} = \lambda_{23} = 10$, and $\lambda_{31} = \lambda_{32} = 5$. We put the values λ_{ij} in non-increasing order and assign tasks to the positions on particular machines, as presented in Table 3.11. After constructing a schedule on each machine we re-index the tasks according to non-decreasing completion times. We calculate $k^* = \lceil 8(5-2)/(4+5) \rceil = 3$ and set $d^* = C_3 = 8$.

Table 3.11. Assignment of tasks from Example 3.42 to positions in the schedule

Position	1	2	3	4	5	6	7	8
λ_{ij}	λ_{21}	λ_{22}	λ_{23}	λ_{11}	λ_{12}	λ_{13}	λ_{31}	λ_{32}
Processing time p_i	2	3	4	6	7	8	9	11
Completion time C_i	8	10	12	6	7	8	17	21
Scheduling cost	0	10	20	8	4	0	45	65

The cost of the schedule is calculated as the total task earliness/tardiness cost (the last row in Table 3.11) plus the due date penalty 8*2*8, and equals 280.

Cheng in [55] reports a computational experiment showing that for small instances the relative deviation of the solutions obtained by Algorithm 3.41 from the optimal solutions does not exceed 6%. The optimal solutions are obtained by solving the following mixed zero-one programming problem.

$$\min \quad \sum_{i=1}^{n} (\alpha e_i + \beta t_i + \gamma d) \tag{3.47}$$

$$\text{s.t.} \quad C_i - d = t_i - e_i, \; i = 1, \dots, n \tag{3.48}$$

$$C_i = \sum_{j=1}^{m} \sum_{k=1}^{n} C_{jk} x_{ijk}, \; i = 1, \dots, n \tag{3.49}$$

$$C_{jk} = C_{j,k-1} + \sum_{i=1}^{n} t_i x_{ijk}, \; i = 1, \dots, n, j = 1, \dots, m \tag{3.50}$$

$$0 \le \sum_{i=1}^{n} x_{ijk} \le 1, \; i = 1, \dots, n, j = 1, \dots, m \tag{3.51}$$

$$\sum_{j=1}^{m}\sum_{k=1}^{n} x_{ijk} = 1, \; i = 1, \ldots, n \tag{3.52}$$

$$x_{ijk} \in \{0, 1\} \tag{3.53}$$

$$C_{j0} = 0 \tag{3.54}$$

In the above formulation x_{ijk} is a decision variable equal to 1 if task i is scheduled in position j on machine k, and 0 otherwise, C_i is the completion time of task i, and C_{jk} is the completion time of the task scheduled in position j on machine k. The first constraint relates the values C_i, e_i, and t_i according to the definitions of earliness (2.6) and tardiness (2.7). The second constraint allows to calculate the completion time of task i depending on its assigned position, and the third constraint takes into account processing times of tasks to calculate the completion times of consecutive tasks on particular machines. The remaining conditions guarantee that only one task is assigned to each position and that processing starts at time zero on each machine. In practice, only small instances can be solved using the zero-one programming approach.

Adamopoulos and Pappis in [5] propose a heuristic algorithm to minimize function (3.41) in a system of unrelated parallel machines. The position weights are calculated as in Algorithm 3.41. The sequence of tasks is obtained as follows. For each task i, $i = 1, \ldots, n$, we calculate the difference between two shortest processing times of task i, i.e. $\Delta_i = |p_{ix} - p_{iy}|$, where x and y are the indices of machines with the shortest processing time of task i. The tasks are indexed according to non-increasing order of Δ_i. The algorithm first assigns positions to tasks for which Δ_i is large and tries to assign them to machines with the shortest processing time. It assigns at most $k = \lceil n/m \rceil$ tasks to each machine in this way. The remaining tasks are assigned one by one to the machine with the lowest total processing time of the assigned tasks.

Algorithm 3.43 (Adamopoulos and Pappis [5]).

1. Index tasks according to non-increasing order of $\Delta_i = |p_{ix} - p_{iy}|$.
2. Set $k = \lceil n/m \rceil$.
3. Set $i = 1$, $Q_j = \emptyset$, and $n_j = 0$ for $j = 1, \ldots, m$.
4. Find j^* such that $p_{ij^*} = \min_j\{p_{ij}\}$. If $n_j^* < k$ then assign task i to machine j^*, i.e. set $Q_{j*} = Q_{j*} \cup \{i\}$ and set $j^* = j^* + 1$, else assign task i to set Q_0.
5. If $i < n$ then set $i = i + 1$ and go to step 4.
6. Calculate $S_j = \sum_{i \in Q_j} p_{ij}$ for each $j = 1, \ldots, m$.

7. Find machine j^* such that $S_{j*} = \min_j\{S_j\}$ and assign the first task, denote it l, from set Q_0 to machine j^*, $Q_{j*} = Q_{j*} \cup \{l\}$, $S_{j*} = S_{j*} + p_{lj^*}$, and $Q_0 = Q_0 \setminus \{l\}$.
8. If $Q_0 \neq \emptyset$ go to step 7.
9. Find an optimal schedule of tasks from set Q_j on machine j, $j = 1, \ldots, m$, and re-index the tasks according to non-decreasing completion times C_i, $i = 1, \ldots, n$.
10. Calculate $k^* = \lceil n(\beta - \gamma)/(\alpha + \beta) \rceil$ in order to find the optimal due date given by $d^* = C_{k^*}$. Calculate the total cost of the schedule as follows: $\sum_{j=1}^{m} \sum_{i=1}^{n} (\alpha_i e_{ij} + \beta_i t_{ij}) + \gamma d$.

Let us illustrate Algorithm 3.43 with the following example.

Example 3.44. Let us consider $n = 12$ tasks and $m = 2$ machines. The processing times of tasks are given in Table 3.12, and the cost penalties are $\alpha = 4$, $\beta = 5$, and $\gamma = 2$.

Table 3.12. Data for Example 3.44

Task	1	2	3	4	5	6	7	8	9	10	11	12
p_{i1}	105	22	56	76	65	42	55	98	13	105	24	78
p_{i2}	44	72	108	44	56	97	42	55	24	10	117	43
Δ_i	61	50	52	32	9	55	13	43	11	95	93	35

The number of tasks assigned to each machine is $k = 6$. In step 3 of Algorithm 3.43 we create three sets: $Q_1 = \{11, 6, 3, 2, 9\}$, $Q_2 = \{10, 1, 8, 12, 4, 7\}$, and $Q_0 = \{5\}$. In step 4 we assign task 5 to set Q_1. The optimal schedules on both machines are defined by completion times of tasks. On machine 1 we obtain $C_5 = 65$, $C_3 = 121$, $C_9 = 134$, $C_2 = 156$, $C_{11} = 180$, and $C_6 = 222$, and on machine 2 we have $C_8 = 55$, $C_4 = 99$, $C_{10} = 109$, $C_7 = 151$, $C_{12} = 194$, and $C_4 = 238$. The optimal due date equals $d^* = C_4 = 109$, and the schedule cost equals 5668.

A small computational experiment is reported in [5], where the solutions obtained by Algorithm 3.43 are compared with optimal solutions for instances with less than 10 tasks and with the best solution from a million of randomly generated V-shaped schedules for instances with 10 to 16 tasks. Obviously, the computational time required by Algorithm 3.43 is much shorter than time needed by full enumeration. The deviation from optimal solutions is less that 9%, and for majority of instances Algorithm 3.43 found better solution than the random sampling.

Another scheduling problem with a controllable due date is proposed by Chen in [47, 48]. He considers the situation where tasks are supposed to be delivered in batches. All early tasks are delivered in one batch without any additional cost. Each batch of tasks delivered after the due date incurs additional cost θ, independent of the number of tasks in the batch. We denote by D_i the delivery date of task i, $i = 1, \dots, n$. The delivery date is common for all tasks belonging to the same batch. The earliness is equal to $e_i = D_i - C_i$, if $C_i \leq d$, and tardiness to $t_i = D_i - d$, if $C_i > d$. The objective is to minimize the following function:

$$\sum_{i=1}^{n} (\alpha e_i + \beta t_i) + \gamma d + \theta q \tag{3.55}$$

where q is the number of tardy batches. Chen in [47] develops a dynamic programming algorithm of complexity $O(n^5)$ to solve this problem under the assumption that $\alpha \leq \beta$.

Cheng in [58] examines the due date assignment problem with the objective to minimize the following function:

$$\sum_{i=1}^{n} wU_i + \gamma d \tag{3.56}$$

where $U_i = 1$ if task i is late and $U_i = 0$, otherwise $i = 1, \dots, n$. Cheng shows that the optimal schedule for this problem is found by scheduling tasks in the SPT order and assigning the due date so that it coincides with the completion time of task k, where $p_k \leq w/n\gamma \leq p_{k+1}$.

A simple generalization of function (3.41) is to consider task dependent costs as follows:

$$\sum_{i=1}^{n} (\alpha_i e_i + \beta_i t_i + \gamma_i d) \tag{3.57}$$

First Quaddus in [209], and then Baker and Scudder in [16] showed that for a given sequence of tasks on a single machine it is easy to find a due date which minimizes function (3.57). Namely, they show that the first task in the sequence for which $\sum_{i=1}^{j} (\alpha_i + \beta_i) \geq \sum_{j+1}^{n} (\beta_i - \gamma_i)$ is completed at the due date. The problem with arbitrary weights is also considered in [51, 52] and [25]. Generalization of the reasoning given by Bagchi et al. in [15] shows that also in this case Property 3.17 holds, i.e. the optimal schedule is V-shaped. Since the case of the problem with $\gamma = 0$ is NP-hard (see [117]), the problem of minimizing function (3.57) is NP-hard with restrictive and unrestrictive due date and symmetric weights.

3.1.6 Controllable processing times

In many practical situations the processing time of a task may be decreased to some limit at an additional cost. We then talk about controllable processing times. The idea of controllable processing times has been taken from the area of project scheduling. Problems with controllable processing times are considered e.g. in [61, 62, 66, 65, 11, 197]. A survey of early results for scheduling problems with controllable processing times can be found in [199]. Scheduling problems with due dates and controllable processing times were examined by Zdrzałka in [258] and Panwalkar and Rajagopalan in [204].

Panwalkar and Rajagopalan [204] consider the WSAD problem with controllable processing times. Let us assume the regular processing time p_i of task i, $i = 1, \ldots, n$, can be reduced by x_i time units, $0 \leq x_i \leq (p_i - p_i^{min})$, where p_i^{min} is the minimum processing time of task i. The cost of reducing the processing time of task i equals λ_i per time unit. Thus, if task i is compressed by x_i time units, its processing cost equals $x_i \lambda_i$, $i = 1, \ldots, n$. The objective is to find a set of processing times and the sequence of tasks to minimize the cost function (3.58).

$$\sum_{i=1}^{n} (\alpha e_i + \beta t_i + \lambda_i x_i) \tag{3.58}$$

Properties 3.1, 3.2, and 3.11 hold for this problem. Moreover, the following Property 3.22 is formulated in [204].

Property 3.22 *There exists an optimal sequence such that no tasks will be partially compressed, i.e. there exists an optimal schedule such that either $x_i = 0$ or $x_i = p_i - p_i^{min}$.*

Panwalkar et al. in [204] formulate the problem of minimizing function (3.58) as an assignment problem. The matrix of costs c_{ij} incurred by scheduling task i in position j, $i = 1, \ldots, n$, $j = 1, \ldots, n$, is calculated as follows:

$$c_{ij} = \begin{cases} p_i^{min} \xi_j + (p_i - p_i^{min}) \lambda_i & \text{if } \lambda_j \leq \xi_j \\ p_i \xi_j & \text{if } \lambda_j > \xi_j \end{cases} \tag{3.59}$$

where

$$\xi_j = \min\{(j-1)\alpha, (n+1-j)\beta\} \tag{3.60}$$

The corresponding assignment problem is formulated as follows.

Problem 3.45 (Assignment Problem).
Minimize

$$\sum_{i=1}^{n}\sum_{j=1}^{n} c_{ij}y_{ij} \tag{3.61}$$

subject to

$$\sum_{i=1}^{n} y_{ij} = 1, \quad j = 1, \ldots, n \tag{3.62}$$

$$\sum_{j=1}^{n} y_{ij} = 1, \quad i = 1, \ldots, n \tag{3.63}$$

The assignment problem can be solved in $O(n^3)$ time (see [206]).

Algorithm 3.46 (Panwalkar and Rajagopalan [204]).

1. Set $\xi_i = \min(\alpha(1-i), \beta(n+1-i))$, $i = 1, \ldots, n$.
2. Set $k = \lceil n\beta/(\alpha+\beta)\rceil$.
3. If $\lambda_i \le \xi_j$ then set $c_{ij} = p_i^{min}\xi_j + (p_i - p_i^{min})\lambda_i$, else set $c_{ij} = p_i\xi_i$, $i = 1, \ldots, n$, $j = 1, \ldots, n$.
4. Solve the assignment problem with cost matrix c_{ij}.
5. Set the optimal due date equal to the completion time of the task in position k.

Let us illustrate the Algorithm 3.46 with an example.

Example 3.47. Let us consider $n = 5$ tasks with $\alpha = 4$, and $\beta = 5$. Processing times (minimum and regular) and compression costs are given in Table 3.13. The cost matrix is given in Table 3.14. The optimal solution is printed in boldface. The optimal sequence corresponding to this solution is $\{5, 1, 2, 3, 4\}$, where tasks 2, 3 and 4 are compressed, and the smallest due date for which this sequence is optimal equals 16 (three tasks are completed by the due date).

Table 3.13. Data for Example 3.47

Task i	λ_i	p_i^{min}	p_i
1	11	2	3
2	8	2	4
3	10	4	7
4	5	5	9
5	9	3	11

Table 3.14. Cost matrix c_{ij} for Example 3.47

Position j	1	2	3	4	5
ξ_j	0	4	8	10	5
1	0	**12**	88	110	55
2	0	16	**32**	36	40
3	0	28	80	**70**	50
4	0	36	60	70	**45**
5	**0**	44	72	102	55

This problem was generalized by Cheng et al. [67] to the due date assignment problem with the following objective function:

$$\sum_{i=1}^{n} (\alpha e_i + \beta t_i + \lambda_i x_i + \gamma d). \tag{3.64}$$

Properties 3.1, 3.2, and 3.11 hold for this problem. Moreover, if $\beta \leq \gamma$ then $d^* = 0$. For any given sequence of tasks with task k completed exactly at the due date, where k is found using Property 3.11, the objective function is given by equation (3.65).

$$\begin{aligned} &\sum_{i=1}^{k} (\alpha(i-1) + \gamma n - \lambda_i)(p_i - x_i) \\ &+ \sum_{i=k+1}^{n} (\beta(n-i+1) - \lambda_i)(p_i - x_i) + \sum_{i=1}^{n} \lambda_i p_i \end{aligned} \tag{3.65}$$

Since $\sum_{i=1}^{n} \lambda_i p_i$ is constant, the objective is to maximize the following function:

$$\sum_{i=1}^{n} \eta_i x_i \tag{3.66}$$

where

$$\eta_i = \begin{cases} \alpha(i-1) + \gamma n - \lambda_i & \text{for } 0 < i \leq k \\ \beta(n-i+1) - \lambda_i & \text{for } k < i \leq n \end{cases} \tag{3.67}$$

Optimal compression can be found as:

$$x_i = \begin{cases} 0 & \text{if } \eta_i < 0 \\ x_i^{'} & \text{if } \eta_i = 0 \\ p_i - p_i^{min} & \text{if } \eta_i > 0 \end{cases} \tag{3.68}$$

where $x_i^{'}$ is an arbitrary value from the interval $[0, p_i - p_i^{min}]$. Since an optimal compression can be found for any given sequence, the problem reduces to a sequencing problem. An optimal sequence can be found by assigning n tasks to n positions in the sequence. If task i, $i = 1, \dots, n$, is scheduled in position j, $j = 1, \dots, n$, it is assigned a weight w_{ij},

$$w_{ij} = \begin{cases} \alpha(j-1) + \gamma n - \lambda_i & \text{for } 0 \leq j \leq k \\ \beta(n-j+1) - \lambda_i & \text{for } k+1 \leq j \leq n \end{cases} \tag{3.69}$$

and its processing time equals

$$p_{ij} = \begin{cases} p_i & \text{if } w_{ij} < 0 \\ p_i^{'} & \text{if } w_{ij} = 0 \\ p_i^{min} & \text{if } w_{ij} > 0 \end{cases} \tag{3.70}$$

where $p_i^{min} \leq p_i^{'} \leq p_i$, $j = 1, \dots, n$. The coefficients c_{ij}, $i = 1, \dots, n$, $j = 1, \dots, n$, are defined where $c_{ij} = w_{ij}p_{ij}$ corresponds to the cost of scheduling task i in position j. We now look for an assignment of tasks to positions minimizing the total cost. To that end Problem (3.45) is solved.

Cheng at al. [67] consider a special case of the problem with controllable processing times, where $\lambda_j = \lambda$ and $x_i = x$, $i = 1, \dots, n$. They show that if processing times of all tasks are jointly reducible, i.e. all processing times are reduced by the same value x, then an optimal schedule can be found in $O(nlogn)$ time with the following algorithm.

Algorithm 3.48 (Cheng et al. [67]).

1. $k := n(\beta - \gamma)/(\alpha + \beta)$;
2. If $k \leq 0$ then set $k = 0$ else set $k = \lceil n(\beta - \gamma)/(\alpha + \beta) \rceil$.
3. Set $\eta_i = n\gamma + (i-1)\alpha - \lambda$, $i = 1, \dots, k$.
4. Set $\eta_i = \beta(n+1-i) - \lambda$, $i = k+1, \dots, n$.
5. Order the weights η_i in non-increasing order.
6. Obtain the optimal sequence by matching the consecutive position weights in descending order with the processing times in ascending order.
7. If $\eta_i > 0$ then $p_i := p_i^{min}$, $i = 1, \dots, n$.
8. Set $d^* = \sum_{i=1}^{k} p_i$.

Biskup and Jahnke [30] examine a more general compression model. They consider the problem of minimizing function (3.14) with jointly reducible processing times. The actual processing time $p_i^{'}$ of task i is proportional to x, i.e. $p_i^{'} = p_i(1-x)$, where p_i is the regular processing time of task i, $i = 1, \dots, n$. Biskup and Jahnke introduce the reduction cost $f(x)$, where $f(x)$ is a monotonous increasing function. It can be proved that Property 3.11 holds for this model. An algorithm, using Algorithm 3.39 in step 1 is proposed to find optimal values of d and x.

Algorithm 3.49 (Biskup and Jahnke [30]).

1. Find an optimal schedule for tasks with regular processing times. Find the optimal cost of this schedule, $F(S^*)$ using Algorithm 3.39.
2. Find the optimal compression x^* of processing times by solving the following program:
   minimize
   $$(1-x)F(S^*) + f(x) \tag{3.71}$$
   subject to
   $$0 \le x \le x_{max} \tag{3.72}$$
3. If $x^* > 0$, reduce the processing times to $p_i^* = (x_i^*)p_i$, $i = 1, \ldots, n$, and calculate the optimal due date d^*.

For the case with linear compression cost function $f(x) = ax$, an optimal sequence exists in which the compression of task processing times is $x = 0$ if $a \le F(S^*)$ or $x = x_{max}$ if $a > F(S^*)$.

Biskup and Cheng in [28] consider the problem of minimization of the following objective function:

$$\sum_{i=1}^{n} (\alpha e_i + \beta p_i + \theta C_i + \lambda_i x_i) \tag{3.73}$$

This problem can be solved optimally in polynomial time by solving a corresponding assignment problem.

Biskup in [27] considers the scheduling problem with the so-called learning effect and the following objective function:

$$\sum_{i=1}^{n} (\alpha e_i + \beta p_i + \theta C_i). \tag{3.74}$$

The learning effect means that the processing time of task i depends on its position in the sequence j according to the relation $p_{ij} = p_i j^k$, where $k \le 0$ is called a learning index. Then the processing time of a task is smaller if the task is scheduled later in the sequence. This problem can be solved in $O(n^3)$ time in two steps. In the first step an assignment problem is solved, and in the second step the optimal due date for the obtained sequence is found.

Cheng et al. in [65] relate the processing time of a task to its start time. The processing time p_i of task i is defined as follows: $p_i = b -$

$aS_i(\sigma)$, where $S_i(\sigma)$ is the start time of task i in schedule σ, and a, b are non-negative integers. Processing time which becomes shorter if a task is started later may be another model of the learning effect. The worker requires less time to complete the tasks started later, because with time he gains knowledge and skills. The objective is to minimize function (3.41). The properties of optimal solutions and an optimization algorithm of complexity $O(nlogn)$ are given in [65].

Alidaee and Ahmadian in [7] consider the problem of minimizing the objective function (3.58) in the system of unrelated parallel machines. A transformation of this problem to the transportation problem which is known to be polynomially solvable (see [206]) is presented in [7].

Chen et al. [49] consider discretely controllable processing times. A finite set of possible processing times and corresponding processing costs are given for each task. The objective is to minimize the total earliness, tardiness and processing cost. For unrestrictive due date and task independent unit earliness and tardiness costs the problem may be transformed to the assignment problem and thus is polynomially solvable.

Ng et al. in [198] consider a common due date assignment problem with processing times dependent on the amount of additional resource allotted to a task. The objective is to minimize a linear combination of scheduling, due date position and resource consumption costs. The scheduling costs are either the earliness/tardiness or the weighted number of tardy tasks. Heuristic algorithms are proposed and tested on a basis of a computational experiment.

3.1.7 Resource dependent ready times

Not only processing times may depend on the amount of a resource allotted to a task. Ventura et al. in [245] consider the earliness/tardiness problem with resource dependent task ready times. Such situation appears, for example, if a task goes first through a preprocessing phase whose length depends on the amount of the allotted resource. Only after completion of the preprocessing phase the task is ready for processing. The problem is formulated as follows. Consider n independent nonpreemptive task with a common due date d, unit earliness cost α and unit tardiness cost β. Let p_i denote processing time and r_i the ready time of task i, $i = 1, \ldots, n$, respectively. It is assumed that the resource consumption of task i is a non-increasing function $f(r)$ of ready time r, defined as follows:

$$f(r) = \begin{cases} a - br & \text{if } \ r \le a/b \\ 0 & \text{if } \ r > a/b \end{cases} \tag{3.75}$$

where a and b are known parameters.

The objective is to minimize the following function:

$$\sum_{i=1}^{n} (f(r_i) + \alpha \max\{0, d - C_i\} + \beta \max\{0, C_i - d\}) \tag{3.76}$$

Several properties of optimal schedules for this problem are proved in [245]. Obviously, Property 3.1 holds. Other properties shared by the problems with a common due date are slightly modified in this case. They are presented below.

Property 3.23 *There exists an optimal schedule for problem (3.76) such that at least one task starts or completes at time d or at time a/b.*

Consider function $g(r) = f(r)+\alpha \max\{0, d-(r+p)\}+\beta \max\{0, (r+p)-d\}$. Function $g(r)$ is piecewise linear and convex. Let us denote by ρ the point at which function $g(r)$ attains its minimum, i.e. $\rho = \min\{r' \ge 0 : g(r') \le g(r) \text{ for all } r \ge 0\}$. Now let us define set of tasks completed by ρ, $E' = \{i : C_i \le \rho\}$ and set of tasks completed after time ρ, $T' = \{i : C_i > \rho\}$. The following property addresses the V-shape of an optimal schedule.

Property 3.24 *There exists an optimal schedule for problem (3.76) such that*

(i) tasks in set E' are arranged in the LPT order;
(ii) tasks in set T' are arranged in the SPT order;
(iii) if there is a task i such that $C_i - p_i \le \rho \le C_i$ and task i is neither the first nor the last task in the schedule, then $p_i \le \max\{p_j, p_k\}$, where j and k are the tasks schedules immediately before and after task i, respectively.

The problem is NP-hard in the ordinary sense. A heuristic and a dynamic programming algorithm are proposed in [245] to solve this problem. The heuristic is developed for three cases:

(i) $a/b \le d$;
(ii) $a/b > d$ and $b \le \beta$;
(iii) $a/b > d$ and $b > \beta$.

In all cases the heuristic first calculates position weights and then assigns the longest task time to the position with the smallest weight, the second longest task to the position with the second smallest weight etc. In the next step optimal completion times of tasks are calculated. Finally an interchange procedure is applied to improve the schedule.

Another algorithm proposed in [245] to solve problem 3.76 is a dynamic programming algorithm. Let us assume that $h_k(s)$ is the optimum cost of scheduling k shortest tasks assuming that the first task starts at time s, whith $\max\{0, \rho - p_1 - p_n\} \leq s \leq \rho + p_n$. The cost of scheduling task k assuming that its processing starts at time s, is equal to $g_k(s) = f(s) + \alpha \max\{0, d - (s + p)\} + \beta \max\{0, (s + p) - d\}$. The recurrence relation is following:

$$h_k(s) = \min\{g_k(s) + h_{k-1}(s + p_k), g_k(s + \sum_{i=1}^{k-1} p_i) + h_{k-1}(s)\} \quad (3.77)$$

The recurrence procedure is initiated with $h_1(s) = g_i(s)$. The optimal solution is found for $\min_s\{h_n(s)\}$. The complexity of the dynamic programming algorithm is $O(n \sum_{i=1}^{n} p_i)$.

Computational results reported in [245] show that the dynamic programming algorithm can solve up to 40 tasks if run on a Pentium 166 MHz computer. For the same set of test instances the heuristic find solutions with average relative deviation from optimum equal to 0.014%.

3.1.8 Common due window

Liman et al. [176] consider the weighted problem in which tasks have a common due window $(d, d + D)$, $d \geq 0$, $D \geq 0$. The earliness and tardiness are defined as follows

$$e_i = \max\{0, d - C_i\} \quad (3.78)$$

$$t_i = \max\{0, C_i - d - D\}. \quad (3.79)$$

Algorithm 3.50 of complexity $O(nlogn)$ finds an optimal size and location of the window as well as an optimal sequence which minimizes the cost function

$$\sum_{i=1}^{n} \alpha e_i + \beta t_i + \delta d + \gamma D \quad (3.80)$$

Algorithm 3.50 (Liman et al. [176]).

1. Set position weights $w_j = \min\{n\delta + (j-1)\alpha, n\gamma, (n+1-j)\beta\}$, $j = 1, \ldots, n$.
2. Assign the longest task to the position with the smallest weight, the second longest tasks to the position with the second smallest weight, etc.
3. Set $d = \sum_{i \in E} p_i$, $D = \sum_{i \in J \setminus T} p_i$.

Let us consider the following example.

Example 3.51. Consider the following problem with $n = 7$ tasks, $\alpha = 12$, $\beta = 19$, $\delta = 4$, and $\gamma = 8$. The processing times of tasks are the following: $p_1 = 2$, $p_2 = 3$, $p_3 = 4$, $p_4 = 6$, $p_5 = 8$, $p_6 = 10$, $p_7 = 15$. The position weights are calculated in Table 3.15. The optimal sequence is $\{6, 4, 31, 2, 5, 7\}$, $d^* = 20$, and $D^* = 25$.

Table 3.15. Calculation of the position weights w_j for Example 3.51

Position j	1	2	3	4	5	6	7
$n\delta + (j-1)\alpha$	28	40	52	64	76	88	100
$n\gamma$	56	56	56	56	56	56	56
$(n+1-j)\beta$	133	114	95	76	57	38	19
w_j	28	40	52	56	56	38	19

Yeung et al. in [256] consider minimization of the total earliness and tardiness in a two machine flow shop with a common due window. The problem is shown to be NP-hard. Branch and bound as well as heuristic algorithms are proposed.

Common due window is also considered by Janiak and Winczaszek in [126, 127]. In this case, however, the start time and the length of the due window are decision variables. Yeung at al. in [255] consider the problem where the size of the due window is fixed, but its location is a decision variable. Two optimality criteria are considered. In [254] the problem of minimizing the weighted number of early and tardy tasks and the task flow time is examined. In [255] the objective is to minimize the earliness/tardiness cost, the weighted number of early and tardy tasks and the lead time. Agreeable unit earliness/tardiness costs are assumed. In both cases the problem is NP-hard. Pseudopolynomial time algorithms are proposed to solve these two problems in [254] and [255], respectively.

A wide range of practical applications brings new problems, challenging for the scheduling theory. Therefore there are a great variety of scheduling problems with earliness and tardiness costs expressed by linear functions.

3.2 Quadratic cost function

In some situations it is desirable to penalize large deviation from the due date more severly. One way to to do that is to apply the quadratic function as the earliness and tardiness cost function. The objective is then defined as minimization of the mean square deviation of task completion times from the common due date. In this section we consider the problem of scheduling n independent and non-preemptive tasks with arbitrary processing times on a single machine with the objective to minimize function (3.81).

$$\sum_{i=1}^{n} (e_i^2 + t_i^2) = \sum_{i=1}^{n} (C_i - d)^2. \tag{3.81}$$

We call a scheduling problem with objective function (3.81) the mean square deviation problem (MSD).

Obviously, there is no idle time inserted between tasks in an optimal schedule, i.e. Property 3.1 holds for the MSD problem. Bagchi et al. in [15], and De et al. in [81] prove that an optimal schedule has a V-shape.

Property 3.25 *In a schedule minimizing the mean square deviation, tasks completed before the shortest task are ordered according to non-increasing processing times, and tasks completed after the shortest task are ordered according to non-decreasing processing times.*

It is worth noticing that the V-shape is now constructed around the shortest task, not as in the case of linear cost functions, around the due date. The V-shape property of schedules with total quadratic deviation from a common due date is discussed in [190]. Sometimes it is possible to find the point t around which the optimal sequence is V-shaped, as proposed by Al-Turki et al. in [6]. Namely, for the objective function (3.82), which reduces to MSD for $v = 1$ and $w = 0$, Theorem 3.52 gives the point t in time around which the optimal sequence is V-shaped.

$$\sum_{i=1}^{n} [(vC_i - d)^2 + wC_i] \tag{3.82}$$

Theorem 3.52 ([6]). *There exists an optimal sequence for the objective function (3.82), which is V-shaped about* $t = d/v - w/(2v^2)$. *Furthermore, if* $d \leq vp_s + w/(2v)$, *where* p_s *is the processing time of the shortest task, then ordering tasks according to the non-decreasing processing times gives an optimal sequence.*

It is easy to notice that for the problem with objective function (3.81) we have $t = d$.

Similarly as for the linear earliness/tardiness cost functions some properties of optimal schedules as well as solution procedures depend on the position of the due date. Figure 3.13 shows the mean square deviation as a function of d for three different schedules obtained for a given set of tasks. The solid line indicates the optimal MSD profile. The minimum MSD value for a given schedule is obtained for $d = \frac{1}{n}\sum_{i=1}^{n} C_i$, i.e. for the due date equal to the mean completion time of all tasks. It is easy to notice from Figure 3.13 that increasing the due date may lead to better solutions.

Property 3.26 *For any given instance of the MSD problem the objective function is a non-increasing function of the common due date.*

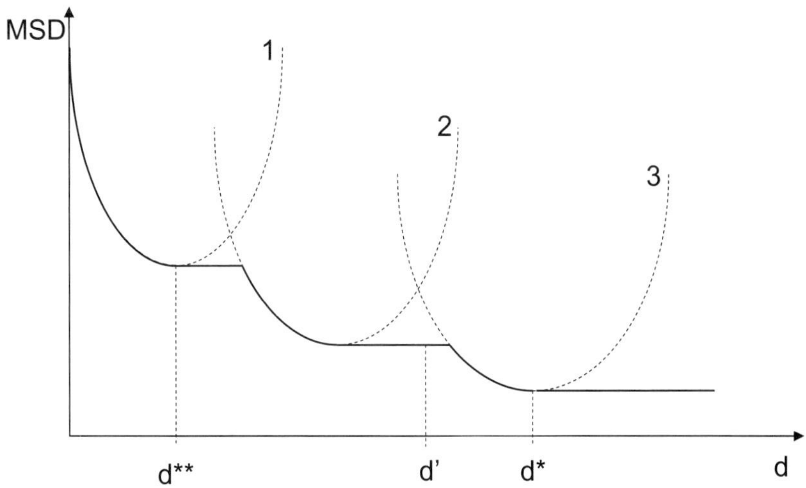

Fig. 3.13. Optimal MDS profile

We denote by d^* the global optimum of the mean square deviation, and by d^{**} the smallest value of d at which a local minimum is obtained. It is worth noticing that the optimal value of MSD at d' is not obtained from schedule 3, but rather from schedule 2 translated in such a way that its minimum coincides with d'. Following Bagchi et al. (see [15]) and De et al. (see [79]), let us define

$$d^{**} = \max\{d : \sum_{i=1}^{n} (C_i^* - d_1)^2 > \sum_{j=1}^{n} (C_j^* - d_2)^2, d_1 < d_2 \le d\} \quad (3.83)$$

and

$$d^{*} = \min\{d : \sum_{i=1}^{n} (C_i^* - d_1)^2 = \sum_{j=1}^{n} (C_j^* - d_2)^2, d_1 \ne d_2, d_2 \le d\} \quad (3.84)$$

We call the problem of minimization of the mean square deviation from a common due date *tightly restricted* if $d \le d^{**}$, *restricted* if $d^{**} < d < d^*$, and *unrestricted* if $d \ge d^*$.

In Section 3.2.1 we consider the unrestricted MSD problem, and in Section 3.2.2 the restricted and tightly restricted problems. Finally, in Section 3.2.3 we present other models, among others problems involving two optimization criteria or problems with multiple machines.

3.2.1 Completion Time Variance

Bagchi et al. in [14] show that if the MSD problem is unrestricted, then the minimization of function (3.81) is equivalent to minimization of the completion time variance (CTV), i.e.

$$\sum_{i=1}^{n} \left(C_i - \frac{1}{n} \sum_{i=1}^{n} C_i \right)^2 . \quad (3.85)$$

The CTV problem was first formulated by Merten and Muller in [185]. The motivation for considering this problem comes from computer file organization problems, where the objective is to minimize the variance of retrieval times for data files requested by the users. Kanet in [141] extends the possible applications to all service systems where approximately the same level of service is desired for all clients. This motivation also applies to scheduling in the just-in-time systems.

Let us index the tasks according to the non-decreasing order of processing times. Schrage in [215] formulates a conjecture about the positions of four longest tasks in an optimal sequence. Namely, he suggests that the optimal sequence has the form $\{n, n-2, n-3, \ldots, n-1\}$. Kanet in [141] shows an instance with 8 tasks for which no optimal schedule has this form. A weaker version of the conjecture in which the positions of three longest tasks are fixed is proved by Hall and Kubiak in [116]. We present it below as Property 3.27.

Property 3.27 *A schedule minimizing the completion time variance on a single machine for a finite set of tasks has the form* $\{n, n-2, \ldots, n-1\}$.

The V-shape property holds for the CTV problem (see [41]). Another important property of optimal schedules for the CTV problem is the following one, proved by De et. al in [81].

Property 3.28 *For a given schedule, the value of the objective function (3.85) does not change if the order of the last* $n-1$ *tasks is reversed.*

Cheng and Kahlbacher in [64] prove a general result saying that if the earliness and tardiness costs are defined by any unimodal function $f(C_i - d)$ then in an optimal schedule the longest task is scheduled first.

Cai in [39] proves that the weighted CTV problem is NP-complete. Kubiak in [155] shows that for $n \geq 3$ the CTV problem is equivalent to the Boolean problem, given by (3.86), and in [153] proves that problem (3.86) is NP-hard. The proof is by reduction from the partition problem. Concluding, the CTV problem is NP-hard.

$$\max \sum_{2 \leq j < i \leq n} w_i v_j x_i \oplus x_j \tag{3.86}$$

where

$$x_i = \begin{cases} 1 & \text{if task } i \text{ is scheduled after task } 1, \\ 0 & \text{otherwise} \end{cases} \tag{3.87}$$

and

$$x_i \oplus x_j = (1 - x_i)x_j + (1 - x_j)x_i, \ 2 \leq j < i \leq n, \tag{3.88}$$

$$w_i = (n - i + 2)p_i + \sum_{j=1}^{i-1} p_j, \ 1 \leq i \leq n \tag{3.89}$$

$$v_j = (j - 1)p_j - \sum_{i=1}^{j-1} p_i, \ 2 \leq j \leq n \tag{3.90}$$

$$v_1 = 0 \tag{3.91}$$

Pseudopolynomial time dynamic programming algorithms to solve the CTV problem are proposed by De et al. in [82], Kubiak in [155],

Kahlbacher in [137] and Jurisch et al. in [136]. In [136] the CTV problem is tranformed to the MinClique problem (Problem 3.30) and solved using the dynamic programming algorithm presented in Section 3.1.3.

The two dynamic programming algorithms proposed by Kubiak in [155] solve in fact the equivalent problem (3.86). Below we present both the algorithms. In the first algorithm let $J(k,w)$ be the set of all schedules in which the sum of weights w_i of tasks $2,\ldots,k-1$, scheduled after the shortest task equals w, i.e. $J(k,w) = \{\mathbf{y} \in \{0,1\}^n : \sum_{i=2}^{k-1} w_i y_i = w\}$. We consider only vectors $\mathbf{y}$ with $y_1 = 0$. Let us define

$$h_k(\mathbf{y}) = \sum_{i=k}^{n} \sum_{j=2}^{i-1} w_i v_j x_i \oplus x_j, k = 3,\ldots,n \tag{3.92}$$

and

$$h(k,w) = \max_{\mathbf{y}\in J(k,w)} \{h_k(\mathbf{y})\}. \tag{3.93}$$

We obtain the following recurrence relation:

$$\begin{aligned} h(k,w) = \max\Bigg\{ & h(k+1,w) + v_k w, \\ & h(k+1,w+w_k) + v_k \left(\sum_{i=2}^{k-1} w_i - w\right)\Bigg\} \end{aligned} \tag{3.94}$$

The initial condition is $h(n+1,w) = 0$ for all w. The solution is found for $h(3,0)$. The solution can be obtained in time $O(n\sum_{i=1}^{n-1}(2i-n)p_i)$.

The second, symmetric algorithm is constructed as follows. We define

$$g_k(\mathbf{y}) = \sum_{j=2}^{k} \sum_{i=j+1}^{n} w_i v_j x_i \oplus x_j, k = 2,\ldots,n-1 \tag{3.95}$$

and

$$g(k,v) = \max_{\mathbf{y}\in I(k,v)} \{g_k(\mathbf{y})\} \tag{3.96}$$

where $I(k,v) = \{\mathbf{y} \in \{0,1\}^n : \sum_{i=k+1}^{n} v_i y_i = v\}$.

We obtain the following recurrence relation:

$$g(k,v) = \max\left\{g(k-1,v) + vw_k,\right.$$
$$\left. g(k-1, v+v_k) + w_k\left(\sum_{i=k+1}^{n} v_i - v\right)\right\} \qquad (3.97)$$

The initial condition is $g(1,v) = 0$ for all v. The solution is found for $g(n-1,0)$. The solution can be obtained in time $O(n\sum_{i=3}^{n} 2(n-i+1)p_i + (n-2)(p_1+p_2))$.

Let us illustrate the latter dynamic programming algorithm with a numerical example.

Example 3.53. Let us consider $n = 7$ tasks with processing times $p_1 = 2$, $p_2 = 4$, $p_3 = 5$, $p_4 = 8$, $p_5 = 21$, $p_6 = 67$, $p_7 = 85$. According to Property 3.27, task 7 is scheduled first, so we skip it in our dynamic program. We calculate $\sum_{i=1}^{5}(2i-6)p_i = 84$, $\sum_{i=3}^{6} 2(7-i)p_i + 4(2+4) = 330$ and

$v_2 = 2 \quad w_3 = 31$
$v_3 = 4 \quad w_4 = 43$
$v_4 = 13 \quad w_5 = 82$
$v_5 = 65 \quad w_6 = 174$

The optimal solution of problem (3.86) is $x_6 = 1$, $x_1 = x_2 = x_3 = x_4 = x_5 = 0$, and $g(6,0) = 14616$. Sequence $\{7,5,4,3,2,1,6\}$ is optimal.

The dynamic programming algorithm proposed by De et al. in [82, 83] uses the following property of optimal V-shaped schedules. Assume that we construct the optimal schedule by successively adding longer tasks.

Property 3.29 *Let S_k and $S_k^{'}$ be V-shaped partial schedules, consisting of k tasks each, with completion times C_i, and $C_i^{'}$ $i = 1,\dots,n$, respectively. Let $\sum_{i=1}^{k} C_i = \sum_{i=1}^{k} C_i^{'}$ and $\sum_{i=1}^{k}(C_i - \frac{1}{k}\sum_{j=1}^{k} C_j) < \sum_{i=1}^{k}(C_i^{'} - \frac{1}{k}\sum_{j=1}^{k} C_j^{'})$. Then sequence S' cannot yield an optimal schedule on completion.*

According to Property 3.29, a simple rule may reduce significantly the computational effort of the dynamic programming algorithm.

Rule 1. Given partial schedules with the same mean completion time choose the one with the smallest variance.

Another property proved by De et al. in [82] is the following.

Property 3.30 *There is an optimal sequence S^* for the CTV problem, such that*

(a) $\frac{1}{n}\sum_{i=1}^{n} C_i^* \le \frac{1}{2}(\sum_{i=1}^{n} p_i + p_n)$, *and*

(b) $\frac{1}{n}\sum_{i=1}^{n} C_i^* \ge \frac{1}{2}\sum_{i=1}^{n} p_i$.

The following rule, further reducing the complexity of the dynamic programming algorithm, is based on Property 3.30.

Rule 2. Keep a partial sequence of k tasks for further consideration only if it fulfils the following conditions:

(a) $\sum_{i=1}^{k} C_i \le \frac{n}{2k}(\sum_{i=1}^{n} p_i - p_n) - (n-k-1)\sum_{i=1}^{k} p_i - \sum_{i=k+1}^{n} (n-i)p_i$, and

(b) $\sum_{i=1}^{n} C_i \ge \frac{n}{2k}\sum_{i=1}^{n} p_i - \sum_{i=k+1}^{n} ip_i$.

The dynamic programming algorithm based on Rules 1 and 2 is presented below.

Algorithm 3.54 (De et al. [82]).

1. Start with a sequence in which only the shortest task is scheduled.
2. For $k = 2$ to n construct two partial schedules: one with task k scheduled at the front, and one with task k scheduled at the end of S_{k-1}. Use Rule 1 and Rule 2 to select nondominated sequences.
3. At the end of stage n select the optimal sequence.

Obviously, the computational time required by the algorithm may be further decreased by scheduling the three longest tasks according to Property 3.27, and applying the dynamic programming algorithm to the remaining $n-3$ tasks. The complexity of Algorithm 3.54 is $O(n^2\sum_{i=1}^{n} p_i)$. Computational experiments reported in [82] show that Algorithm 3.54 solves instances up to 100 tasks with processing times generated randomly from the interval $[1, 100]$ in less than 80 seconds on a VAX8600 computer.

De et al. in [82] and Cai in [40] develop fully polynomial time approximation schemes for the CTV problem. The approximation scheme proposed by De et al. in [82] is based on the same idea as Algorithm 3.54. Assume that $n-1$ tasks are scheduled using Algorithm 3.54. We denote by $\overline{C}^L$ and $\overline{C}^H$, respectively, the minimum and maximum value of the mean completion time of all schedules yielded by the algorithm in the first $n-1$ stages. We partition the interval $[\overline{C}^L, \overline{C}^H]$ into subintervals with $\Delta = (\epsilon LB)/(2n\sum_{i=1}^{n-1} p_i)$, where ϵ is the acceptable relative error and LB is a lower bound on the completion time variance. The lower bound LB is calculated as an optimal solution to the following relaxed CTV problem:

$$\text{minimize} \quad \mathbf{x}^T D \mathbf{x}^T \tag{3.98}$$

$$\text{subject to} \quad \mathbf{1}^T \mathbf{x} = \sum_{i=1}^{n-1} p_i \tag{3.99}$$

$$\mathbf{a} \leq \mathbf{x} \leq \mathbf{b} \tag{3.100}$$

where $\mathbf{x}$, $\mathbf{a}$, and $\mathbf{b}$ are $(n-1)$ column vectors of variables and lower and upper limits of the variable values, respectively. The lower and upper limits for the variable values follow from simple observations that

$$x_1 = p_n, x_i \geq p_1, \ i = 2, \ldots, n$$
$$x_i, x_{n-i+1} \leq p_{n-j+1}, \ i = 2, \ldots, \lceil (n+1)/2 \rceil.$$

Moreover, D is a $(n-1) \times (n-1)$ symmetric matrix whose k-th row is given by

$$(1/n^2)[(n-k), 2(n-k), \ldots, k(n-k), (n-k-1)k, \ldots, 2k, k]$$

Problem (3.98) can be solved iteratively in $O(n)$ time as follows.

$$x_j = x_{n-j+2}$$
$$= \min\left\{p_{n-j+1}, p_1 + \max\left\{0, \frac{1}{2}\left[\sum_{i=1}^{n-1} p_i - (n-1)p_i\right.\right.\right.$$
$$\left.\left.\left. - 2\sum_{k=2}^{j-1} p_{n-k+1}\right]\right\}\right\}$$

for $j = 2, \ldots, \lceil (n+1)/2 \rceil$.

The following Rule 3 is applied instead of Rules 1 and 2 in Algorithm 3.54 and in the approximation scheme.

Rule 3. Given all partial schedules with mean completion time in a certain interval, choose the one with the smallest variance.

The complexity of the approximation scheme is $O(n^3/\epsilon)$. Computational experiments reported in [82] show that the relative error of solutions generated by the approximation scheme is significantly less than ϵ.

Ng et al. in [196] prove the existence a tight lower bound on the CTV value. We present it below, together with the upper bound proved by De et al. in [82].

Theorem 3.55 ([196, 82]). *There exists an optimal sequence for the CTV problem, such that*

$$\frac{\sum_{i=1}^{n} p_i}{2} + \frac{(n-1)(p_n - p_{n-1}) + \sum_{i=1}^{n} C_i}{2n} \leq \frac{1}{n}\sum_{i=1}^{n} p_i$$

and

$$\frac{1}{n}\sum_{i=1}^{n} C_i \geq \frac{\sum_{i=1}^{n} p_i + p_n}{2}$$

for $n \geq 2$.

Branch and bound algorithms for the CTV problem were developed by Bagchi et al. in [14], De et al. in [81], and Viswanathkumar and Srinnivasan in [246]. The first two algorithms address the general MSD problem, while the last one is dedicated to the CTV problem. Below we present the branch and bound algorithm for the CTV problem, proposed by Viswanathkumar and Srinnivasan in [246].

We assume that tasks are indexed according to non-decreasing processing times and that the three longest tasks are assigned to positions according to Property 3.27, so the first task starts at time zero, the last task starts at time $\sum_{i=1}^{n} p_i - p_{n-1}$, and the initial partial sequence has the form $\{n, n-2, \ldots, n\}$. Each node corresponds to a partial schedule. Two successors of each node are created. One of them corresponds to the sequence obtained by fixing the consecutive task in the first available position, and the other one to the sequence where the consecutive task is fixed in the last available position. The branching is performed according to the best-first rule. At each node the lower bound is calculated as follows. Let us assume that $l < n$ longest tasks are fixed. We temporarily fix tasks in non-increasing order of processing times in either forward or backward positions until there is one more task fixed in the forward direction is than in the backward direction. If $k \geq l$ tasks are temporarily fixed, the lower bounds are calculated as follows:

$$L_k = V_1/n + [(p_1 + p_2)^2 + (p_1 + \ldots + p_4)^2 + \ldots + (p_1 + \ldots + p_{n-(k+3)})^2]/2n$$

if n is odd, and

$$L_k = V_1/n + [(p_1)^2 + (p_1 + \ldots + p_3)^2 + \ldots + (p_1 + \ldots + p_{n-(k+3)})^2]/2n$$

if n is even.

If at any node the number of temporarily fixed tasks equals $n - 2$, the value of the completion time variance is calculated for both possible sequences. If any result is less than all other lower bounds, then the solution is optimal.

Computational experiments reported in [246] show that the branch and bound algorithm solves instances with up to 30 tasks in less than one hour. The algorithm was compared with the dynamic programming algorithm proposed by De et al. in [82] for instances with up to 15 tasks. The proposed branch and bound algorithm proved to be more efficient. Moreover, if the limit on the maximum number of active nodes is set at 100, the algorithm finds solutions close to the lower bounds for instances with up to 100 tasks in a few seconds. It generates better solutions than the heuristic proposed by Manna and Prasad in [181].

Vani and Raghavachari in [242] propose the following heuristic algorithm for the CTV problem. Starting with any V-shape sequence, the algorithm checks if interchanging the positions of two tasks may improve the value of the objective function. The tasks are fixed sequentially in positions at both ends of the schedule. They use formula (3.101) in order to evaluate the cost of interchanging the task in the $(k + 1)$st position in the sequence with the task in the $(n - l)$th position.

$$\Delta(k, l) = \frac{p_{k+1} - p_{n-l}}{n}[2Y_m - 2(n - m)Z + (n - k - l)(k - l - 1)(p_{k+1} + p_{n-l})] \quad (3.101)$$

where

$$Y_m = \sum_{i=k+2}^{n-l-1} (kn - i(k + l) + k + l)p_i \quad (3.102)$$

and

$$Z = \sum_{i=n-l+1}^{n} (n-i+1)p_i - \sum_{i=1}^{k} (i-1)p_i \tag{3.103}$$

The heuristic is presented below in a more formal way.

Algorithm 3.56 (Vani and Raghavachari [242]).

1. Start with a V-shaped sequence, for example a sequence obtained by Algorithm 3.1.
2. Set $k = 1$, $l = 1$.
3. Calculate Δ according to formula (3.101).
4. If $\Delta \leq 0$ then fix the task in position $k+1$. Set $k = k+1$. If $k + l \leq n - 1$ then go to step 3, else go to step 6.
5. If $\Delta > 0$ then fix the task from position $k+1$ in position $n - l$. Set $l = l + 1$. If $k + l \leq n - 1$ then go to step 3, else go to step 6.
6. Repeat step 2 through step 5 until the values of Δ in step 3 are less than or equal to zero for all tasks.

If $\Delta \leq 0$ for all tasks, then no single interchange reduces the value of the objective function. Algorithm 3.56 finds optimal solutions for 5 out of 7 instances examined by Kanet in [141]. Solutions for the remaining two instances are close to optimal.

Gupta et al. in [111] present a heuristic genetic algorithm for solving the CTV problem. Jurisch et al. in [136] propose heuristics based on solving the MinClique problem, which is equivalent to the CTV problem, using the so-called spectral algorithms. The best solutions obtained by the spectral algorithm are improved by two types of procedures: an interchange procedure and a tabu search algorithm.

Cai in [40] considers the weighted CTV problem with the objective function formulated as follows:

$$\sum_{i=1}^{n} w_i \left(C_i - \frac{1}{w} \sum_{i=1}^{n} w_i C_i \right)^2 \tag{3.104}$$

where $w = \sum_{i=1}^{n} w_i$.

The problem is considered for so-called agreeable weights, i.e. such that $p_i < p_j$ implies $w_i \geq w_j$. It may be proved that in case of agreeable weights an optimal schedule is V-shaped in terms of weighted processing times, meaning that Property 3.12 holds. Cai in [40] proposes a pseudopolynomial time algorithm and a heuristic to solve the weighted CTV problem with agreeable weights. The heuristic finds solutions very close to optimal ones, and in many cases achieves optimal solutions. Moreover, two fully polynomial time approximation schemes are proposed for problems in which weights are bounded by a polynomial of n.

3.2.2 Restricted MSD problem

The restricted and tightly restricted MSD problems do not share all the properties of the CTV problem. For example, it is not always optimal to schedule the longest task first.

Weng and Ventura in [251] observe that d^{**} does not have to be less that d^* for every instance of the MSD problem. In other words, if $d^{**} = d^*$ then the restricted version vanishes.

It is clear from Figure 3.13 that for $d < d^{**}$ the optimum MSD cannot be obtained by translation of a schedule. Therefore the first task has to start at time zero. It is expressed in the following property, given in [81, 251].

Property 3.31 *For any instance of the tightly restricted problem of minimization of the square deviation from a common due date, the start time of the first task in an optimal schedule has to be zero.*

There is no efficient method for determining d^{**} for a given instance, but the next property, proved in [79], provides a lower bound on d^{**}.

Property 3.32 *Let $\overline{C}_{SPT}$ be the mean completion time of the SPT sequence, and*

$$m = \max\left\{ j : \sum_{i=1}^{j} p_i < \overline{C}_{SPT} \right\}.$$

Let $\overline{C}$ be the mean completion time of the sequence $\{m-1, m-2, \ldots, 1, m, \ldots, n\}$. Then $d^{**} \geq \overline{C}$.

Algorithm 3.54 can be adjusted to find optimal solutions for the restricted MSD problem. Weng and Ventura in [251] focus on problems of minimizing the square deviation from a small common due date. They prove several dominance properties and propose a dynamic programming algorithm.

Property 3.33 *Assume that tasks are ordered according to non-decreasing processing times, S^* is an optimal sequence for the tightly restricted problem of minimization of the mean variance of completion times from a common due date, and S_k is a partial sequence consisting of k shortest tasks. Then the start time of the first task in the partial sequence S_k cannot exceed $\sum_{k+1}^{n} p_i$.*

Property 3.34 *Let*

$$r = \min \left\{ j : \sum_{i=1}^{j} p_i \geq d \right\}$$

and

$$q = \min \left\{ j : \sum_{i=1}^{j} p_i \geq \sum_{i=1}^{n} p_i - d \right\}$$

Then in any optimal sequence for the tightly restricted MSD problem

$$C_i - p_i \geq d - p_r - \sum_{j=1}^{i} p_j$$

and

$$C_i \leq d + p_q + \sum_{j=1}^{i} p_j$$

Moreover, as stated earlier in this section, the sequence that minimizes the square deviation from a common due date is V-shaped and contains no processor idle time.

The dynamic programming algorithm proposed in [251] for the tightly restricted problem is defined as follows. We denote by $H(k, s)$ the minimum total cost of scheduling k shortest tasks, given that s is the start time of the first task. We obtain the following recurrence relation

$$\begin{aligned} H(k, s) = \min \Big\{ & H(k-1, s+p_k) + (s + p_k - d)^2, \\ & H(k-1, s) + (s + \sum_{i=1}^{k} p_i - d)^2 \Big\} \end{aligned} \tag{3.105}$$

Ties are broken arbitrarily. If the first term is smaller, task k should be scheduled in the front of the partial schedule of $(k-1)$ tasks, otherwise task k should be scheduled at the back of this partial schedule. The initial conditions are formulated below.

$$H(1, s) = (s + p_1 - d)^2,$$

$$\text{for } \max\{0, d - p_r - p_1\} \leq s \leq \min \left\{ \sum_{i=2}^{n} p_i, d + p_q \right\}$$

$$H(k,s) = \infty, \text{ for } k = 1, \ldots, n, \text{ and}$$

$$\min\left\{\sum_{i=k+1}^{n} p_i, d + p_q + \sum_{i=1}^{k-1} p_i\right\} < s < \max\left\{0, d - p_r - \sum_{i=1}^{k} p_i\right\}$$

The optimal solution is found for $H(n,0)$. The computational complexity of the algorithm is $O(n\sum_{i=1}^{n} p_i)$. Computational experiments reported in [251] show that the algorithm solves instances with up to 100 tasks with processing times randomly generated from the interval $[1,100]$ within 2 seconds. This algorithm can be adapted to solve the CTV problem, although additional effort is required to examine possible values of d.

Branch and bound algorithms for the MSD problem are presented by Bagchi et al. in [14], and De et al. in [81]. The algorithm proposed by De et al. in [81] is designed as a general procedure that can be adjusted to solve unrestricted, restricted or tightly restricted MSD problems. Computational experiments show that the branch and bound algorithm examined in [81] is more efficient than the enumerative procedure proposed by Bagchi et al in [14].

Below we present the branch and bound algorithm proposed by De et al. in [81]. The tasks are scheduled in the LPT order. Each node of the branching tree corresponds to a partial schedule of the form $\{E, \ldots, T\}$, where E and T are fixed sequences of tasks scheduled before and after the due date, respectively, and preserving the V-shape. We denote by S the set of tasks scheduled in the partial schedule, by V the set of tasks remaining to be scheduled, and by C_E the completion time of the last task in set E, i.e. $C_E = \sum_{i\in E} p_i$. The completion time variance of the fixed tasks (in set S) is denoted by $Z^P(S)$, and the completion time variance of the optimal schedule of tasks in set V is denoted by $Z^*(V)$. We denote by $\overline{C}_L^V$ the lower bound on the mean completion time of tasks in set V (it may be easily calculated as the mean completion time of the SPT sequence). Moreover, let $\overline{C}_U^V$ be the upper bound on the mean completion time of tasks in set V (it may be easily calculated as the mean completion time of the LPT sequence). Finally, let us denote by $\overline{C}^P$ the mean completion time of the partial sequence $\{E, \ldots, T\}$.

We define

$$\overline{C}^V = \begin{cases} \overline{C}_L^V & \text{if} \quad \overline{C}^P < \overline{C}_L^V \\ \overline{C}_U^V & \text{if} \quad \overline{C}^P > \overline{C}_U^V \\ \overline{C}^P & \text{otherwise} \end{cases} \tag{3.106}$$

The lower bound is calculated as

$$LB = Z^P(S) + Z^*(V) + (k/n)(n-k)(\overline{C}^P - \overline{C}^V)^2 \tag{3.107}$$

Now, let us take $\overline{C}^V = \overline{C}_0^V$, where $\overline{C}_0^V$ is the mean completion time of the optimal schedule of set V. The upper bound is calculated as

$$UB = Z^P(S) + Z^*(V) + (k/n)(n-k)(\overline{C}^P - \overline{C}^V)^2 \tag{3.108}$$

It is easy to notice that the bounds are calculated recursively. Depending on the problem type (tightly restricted, restricted or unrestricted) the algorithm runs in a slightly different mode. The modes differ in the way the initial upper bound is found. The computational times for the tightly restricted MSD problem are longer than for the unrestricted problem. The best performance of the algorithm is observed for the restricted problem. Instances with up to 40 tasks can be solved with average computational time of 8 hours for the tightly restricted problem. Nevertheless, the algorithm is more efficient than the enumeration procedure proposed by Bagchi et al. in [14].

Bagchi et al. in [15] provide an enumeration scheme based on properties of optimal schedules for the weighted MSD problem, i.e. the problem with the objective function (3.109).

$$\sum_{i=1}^{n} w_i(C_i - d)^2 \tag{3.109}$$

Both unconstrained and constrained versions are considered.

3.2.3 Other models

The problem of minimizing the completion time variance in the system of m parallel identical machine is considered by Cai and Cheng in [43]. The problem is NP-complete in the strong sense when m is arbitrary, and NP-complete in the ordinary sense when m is fixed. Two algorithms are proposed. The first one generates an optimal solution in time $O(n^{2m}P^m(P-P_m)^{m-1}/[m^m(m-1)!]^2)$, where P is the sum of all the processing times and P_m is the sum of the first m largest processing times. The second algorithm can find a near-optimal solution in time $O(nP^m(P-P_m)^{m-1}/m^m(m-1)!)$. It is also shown in [43] that the relative error of the near-optimal solution is guaranteed to approach zero at a rate $O(n^{-2})$ as n increases.

Bagchi in [12] examines the bicriteria problem of minimizing the mean and variance of completion times. Also De et al. extend their results presented in [82] to a bicriteria problem where the objectives are to minimize the mean completion time and the variance of completion times simultaneously. They propose an algorithm constructing a minimal set of solutions optimal for the objective function (3.110) in $O(n^2\sum_{i=1}^{n}p_i)$ time:

$$\eta\sum_{i=1}^{n}C_i+(1-\eta)(\sum_{i=1}^{n}(C_i-\sum_{i=1}^{n}C_i)^2) \tag{3.110}$$

where $0\le\eta\le 1$.

Gupta and Sen propose a more general model, where each task has a distinct due date d_i and the objective is to minimize function (3.111).

$$\sum_{i=1}^{n}(C_i-d_i)^2 \tag{3.111}$$

A branch and bound algorithm and a heuristic are proposed to solve the problem. The branch and bound algorithm solves instances with up to 9 tasks. The heuristic selects only a portion of the best nodes found by the branch and bound algorithm. The reduction of the computational effort is more significant for larger instances.

A problem with similar properties as the CTV problem is the problem of minimizing the waiting time variance. This problem is considered in [12, 185] and [94]. The waiting time is defined as $W_i=C_i-p_i$ and the objective function is the following:

$$\frac{1}{n}\sum_{i=1}^{n}(W_i-\overline{W}) \tag{3.112}$$

where $\overline{W}=(1/n)\sum_{i=1}^{n}W_i$. Merten and Muller in [185] show that the optimal CTV schedule is antithetical to the optimal WTV schedule, i.e. that if a task is assigned to position i in the optimal WTV schedule than it is assigned to position $(n-i+1)$ in the optimal CTV schedule. Eilon and Chowdhurry in [94] show that the optimal WTV schedule is V-shaped.

4

Individual due dates

In this chapter we consider the just-in-time scheduling problems, where each task has a distinct due date. The due dates may be either problem parameters or decision variables.

As we mentioned in Section 2.1.2, in the earliness/tardiness scheduling it may be desirable to delay the completion of some tasks in order to decrease the earliness cost. In the case of distinct due dates such delay may produce machine idle time, as shown in Figure 2.2. However, there may be situations where machine idle time is not allowed, for example if machines are very expensive and should be maximally utilized. The no idle time constraint, although not completely consistent with the just-in-time philosophy, is sometimes considered in the literature. Therefore in this chapter we distinguish two main types of problems: with and without idle time inserted. In general, inserting the idle time leads to better schedules, but makes the problem analysis more difficult. It is worth noticing that in the case of controllable due dates, optimal schedules do not have any idle time inserted. A discussion and experimental investigation of schedules with and without idle time is presented in [142].

Various functions may be used to evaluate the earliness and tardiness costs, but the linear functions are examined most commonly. Therefore in most cases in this chapter we assume that the cost functions are linear.

Very few papers deal with shop systems in the case of individual due dates. Arakawa et al. in [9] propose a simulation based scheduling method to minimize deviations from task due dates in a job shop. They tested this method for data sets extracted from a running factory with 200 tasks and 16 work centers (each work center contains one to six machines) in a single problem instance. Computational experiments performed for 50 test instances show that the proposed method finds better solutions than simple priority rules. The average computational time is 5 minutes on a Pentium II PC, which is reasonable in this type of applications.

In Section 4.1 we present properties of optimal schedules and algorithms for the problem where machine idle time may be inserted, and in Section 4.2 we examine the problem under the assumption that machine idle time is not allowed. The case where due dates are not given in the problem formulation, but are subject to optimization decisions, is examined in Section 4.3.

4.1 Schedules with idle time

In this section we consider the problem of scheduling n independent nonpreemptive tasks on a single machine. The processing time of task i is denoted by p_i, and the due date of task i is denoted by d_i, $i = 1, \ldots, n$. The objective is to minimize function (4.1).

$$\sum_{i=1}^{n} (\alpha_i \max\{0, d_i - C_i\} + \beta_i \max\{0, C_i - d_i\}) \tag{4.1}$$

Garey et al. in [104] prove that the problem is NP-hard even if $\alpha_i = \beta_i = 1$, $i = 1, \ldots, n$. The proof is by reduction from the even-odd partition problem (Problem 3.28). The NP-completeness proof of the even-odd partition problem is also given in [104].

Although the problem with task dependent due dates is NP-hard even for identical weights, some properties of optimal schedules are proved for the special cases with proportional or identical weights α_i and β_i. We start with the presentation of properties of optimal schedules and algorithms developed for the general case (Section 4.1.1), consider the proportional weights in Section 4.1.2, and finally discuss the non-weighted case in Section 4.1.3. In Sections 4.1.4 - 4.1.7 we present other objective functions considered in the case if idle time is allowed.

4.1.1 Arbitrary weights

If the order of tasks is given the scheduling problem with objective function (4.1) can be solved in polynomial time. In such case the optimal completion times of tasks may be found as a solution to the following linear programming problem:

$$\text{minimize} \quad \sum_{i=1}^{n} (\alpha_i e_i + \beta_i t_i) \tag{4.2}$$

subject to

$$C_1 \geq p_1 \tag{4.3}$$
$$C_i \geq C_{i-1} + p_i \quad i = 2, \ldots, n \tag{4.4}$$
$$e_i \geq d_i - C_i \quad i = 1, \ldots, n \tag{4.5}$$
$$t_i \geq C_i - d_i \quad i = 1, \ldots, n \tag{4.6}$$
$$e_i \geq 0 \quad i = 1, \ldots, n \tag{4.7}$$
$$t_i \geq 0 \quad i = 1, \ldots, n \tag{4.8}$$

Although the linear programming problem may be solved in polynomial time, the computational time required to solve instances with many variables and constraints is large. Therefore more efficient algorithms were proposed to solve the scheduling problem with a given sequence of tasks. Garey et al. in [104] propose an algorithm of complexity $O(nlogn)$ for two special cases of the problem, the case with identical weights and the case with identical processing times of all tasks. Chretienne in [69] develops an algorithm solving the problem with arbitrary weights and processing times in $O(n^3logn)$ time. Davis and Kanet in [77, 78] and Szwarc and Muhopadhyay in [235] propose algorithms of complexity $O(n^2)$. Finally, Sourd and Sidhoum in [219] and Bauman and Józefowska in [21] propose two different algorithms, both of complexity $O(nlogn)$, which find optimal schedules for a given sequence of tasks with arbitrary unit costs and processing times.

The procedure proposed in [21] uses some properties of the cost function defined below. Since the sequence of tasks is fixed, it remains only to find completion times of tasks such that the value of function (4.1) is minimal, the first task starts not earlier than at time zero, and the tasks do not overlap. Let us consider an optimal partial schedule S_{k-1}^* of the first $k-1$ tasks, $k = 2, \ldots, n$. Let $C_i(S_{k-1}^*)$, $i = 1, \ldots, k-1$, denote the completion time of task i in schedule S_{k-1}^* and let $K(S_{k-1}^*)$ be the cost of schedule S_{k-1}^*. While scheduling task k, the following two situations have to be considered:

(i) $C_{k-1}(S_{k-1}^*) + p_k \leq d_k$ and
(ii) $C_{k-1}(S_{k-1}^*) + p_k > d_k$.

In case (i), obviously task k may be completed at d_k and the value of the objective function will not increase (see Figure 4.1).

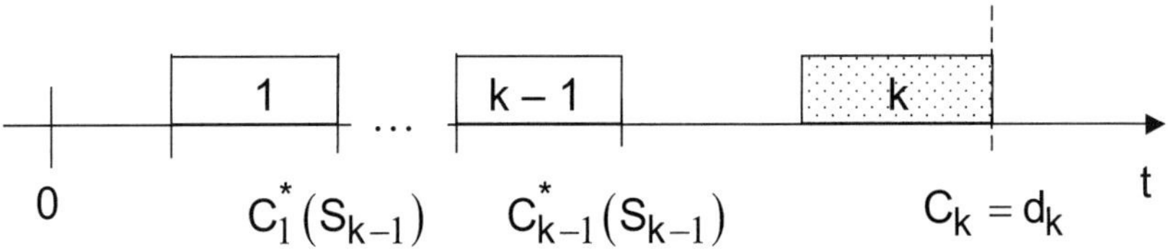

Fig. 4.1. Scheduling the kth task in case (i).

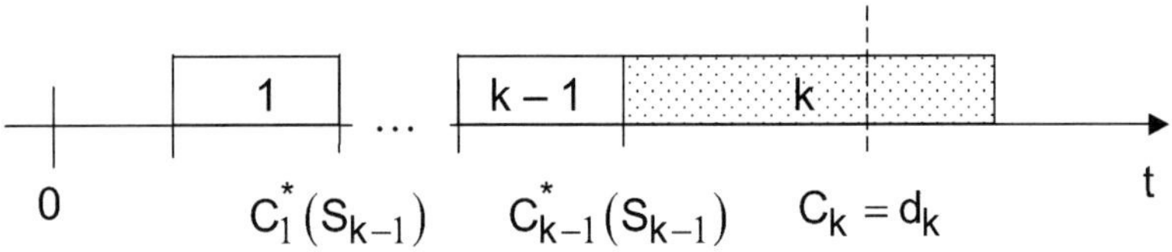

Fig. 4.2. Scheduling the kth task in case (ii).

In case (ii) it has to be decided whether to complete task k after the due date, or to decrease the completion time of task $k-1$ by shifting task $k-1$ and its predecessors to the left by x time units, so that the tardiness of task k decreases by x. By shifting a task we understand a change of task completion time that does not alter the original task permutation. It only makes sense to consider $x \leq C_k - d_k$, because if $x > C_k - d_k$ then task k incurs the earliness cost. The shift never decreases the cost of scheduling tasks $1, \ldots, k-1$. The decision whether task k should be shifted or not is made on the basis of comparison of the tardiness cost of task k, and the additional cost incurred by shifting task $k-1$ to the left. Notice that in order to obtain a feasible schedule, x must not exceed the total idle time in schedule S_{k-1}. Concluding, x must satisfy inequality (4.9).

$$x \leq \min \left\{ C_i - d_i, C_{k-1} - \sum_{i=1}^{k-1} p_i \right\} \tag{4.9}$$

Below we discuss how to calculate the cost incurred by the shift of the last task in any schedule S_{k-1} by x time units to the left. To this end let us first define the procedure LETF-SHIFT(x).

Algorithm 4.1 (LEFT-SHIFT(x) [21]).

1. Set $C_{k-1} = C_{k-1} - x$ and set $i = k - 2$.
2. Set $C_i = \min\{C_i, C_{i+1} - p_{k+1}\}$, else stop.
3. If $k > 1$ then set $k = k - 1$ and go to step 2, else stop.

It is easy to notice that the tasks are shifted until we obtain $C_j \leq C_{j+1} - p_{j+1}$ for some task j. The shift of task k equals $x_k = x$ and we calculate the shift of the remaining tasks according to formula (4.10).

$$x_i = \max\{0, x_{i+1} - (C_{i+1} - C_i - p_{i+1})\}, i = 1, \ldots, k-1 \qquad (4.10)$$

Since the schedule S_k is uniquely determined by the completion times of tasks $1, \ldots, k$, it may be represented by a k-element vector $S_k = [C_1(S_k), \ldots, C_k(S_k)]$. Thus, the schedule obtained by applying the procedure LEFT-SHIFT(x) to S_k is represented by vector $[C_1(S_k) - x_1, C_2(S_k) - x_2, \ldots, C_k(S_k) - x_k]$.

Let us define function $\Delta K_k(x)$ as the difference between the cost of schedule S_k^* and the cost of a schedule obtained by applying procedure LEFT-SHIFT(x) to schedule S_k^*. Thus, when scheduling task k we have to minimize function (4.11).

$$\Delta K_k(x) = \Delta K_{k-1}(x) + \beta_k(C_{k-1} + p_k - x - d_k) + \alpha_k x \qquad (4.11)$$

If task k starts at time C_{k-1}, then the deviation of the completion time of task k from its due date equals $y_k = |C_{k-1} + p_k - d_k|$. We construct function ΔK_k iteratively, starting with the first task. If $d_1 < p_1$, then the first task is started at time zero, and no left shift is possible. If $d_1 \geq p_1$, then the first task completes at time d_1, and its left shift by x time units, $x \leq y_1$, incurs additional cost $\alpha_1 x$, where $0 \leq x \leq d_1 - p_1$. It is not necessary to consider any right shift of the schedule, because such shift never decreases the value of the objective function. Let us assume that the first task is scheduled with $C_1 = d_1$. We proceed to scheduling the second task. In case (i), we construct function $\Delta_2 K(x)$ as shown in Figure 4.3.

It is easy to read from Figure 4.3 that the total idle time in schedule S_2 equals $y_1 + y_2$, so the maximum decrease of the completion time of task 2 is $x \leq y_1 + y_2$. Shifting task 2 by $x \leq y_2$ incurs only earliness cost of task 2, but shifting it by $y_2 \leq x \leq y_1 + y_2$ causes both tasks to be completed early.

If $C_{k-1} + p_k - d_k \leq 0$ then the values of function $\Delta K_k(x)$ can be calculated according to formula (4.12).

$$\Delta K_k(x) = \begin{cases} \alpha_k x & \text{if } 0 \leq x \leq y_k \\ \Delta K_{k-1}(x - y_k) + \alpha_k x & \text{if } y_k \leq x \leq C_{k-1} - \sum_{i-1}^{k-1} p_i \end{cases} \qquad (4.12)$$

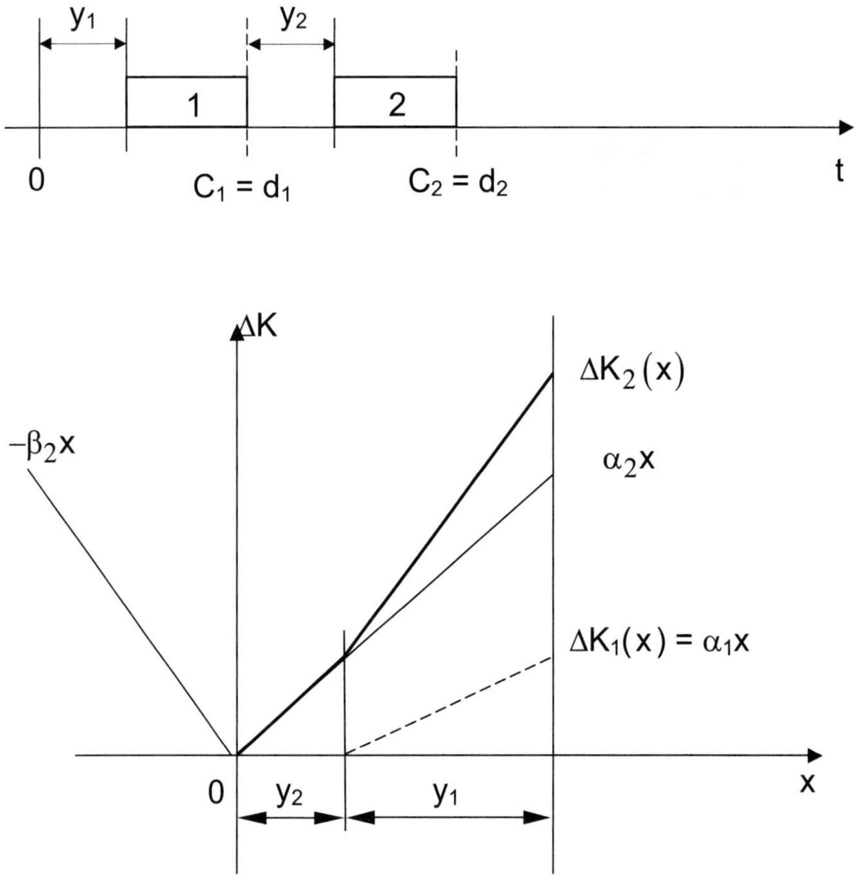

Fig. 4.3. Function ΔK_2 in case (i).

In case (ii) it is not possible to complete task 2 by the due date without shifting task 1 to the left. Such move results in the decrease of the tardiness cost of task 2 and increase of the earliness cost of task 1. Function $\Delta K_2(x)$ is now constructed as shown in Figure 4.4.

If $C_{k-1} + p_k - d_k > 0$ then the values of function $\Delta K_k(x)$ can be calculated according to the following formula:

$$\Delta K_k(x) = \begin{cases} \Delta K_{k-1}(x) + \beta_k(y_k - x) & \text{if } 0 \leq x \leq y_k \\ \Delta K_{k-1}(x) + \alpha_k x & \text{if } y_k \leq x \leq C_{k-1} - \sum_{i-1}^{k-1} p_i \end{cases} \quad (4.13)$$

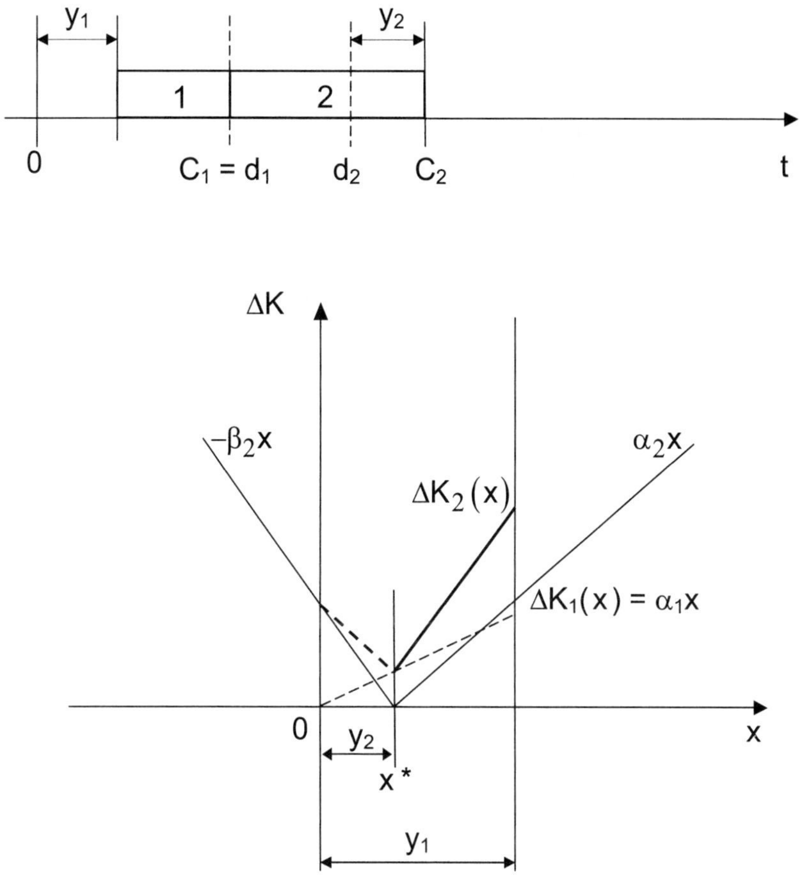

Fig. 4.4. Function ΔK_2 in case *(ii)*.

Proceeding in this way, we define an algorithm to construct function $\Delta K_k(x)$ step by step, as illustrated in Figure 4.5. At each step we find x^* where function $\Delta K_k(x)$ attains its minimum. Since we never shift any task to the right, we simply remove that part of the graph of function $\Delta K_k(x)$ (marked with the bold dashed line) which corresponds to $x < x^*$. It is easy to observe that function $\Delta K_k(x)$ is piecewise linear, convex and increasing for any $k = 1, \ldots, n$. The algorithm scheduling tasks in a given order is based on function $\Delta K_k(x)$. The algorithm is presented below.

Algorithm 4.2 (Bauman and Józefowska [21]).

1. Set $k = 1$ and $C_0^* = 0$.

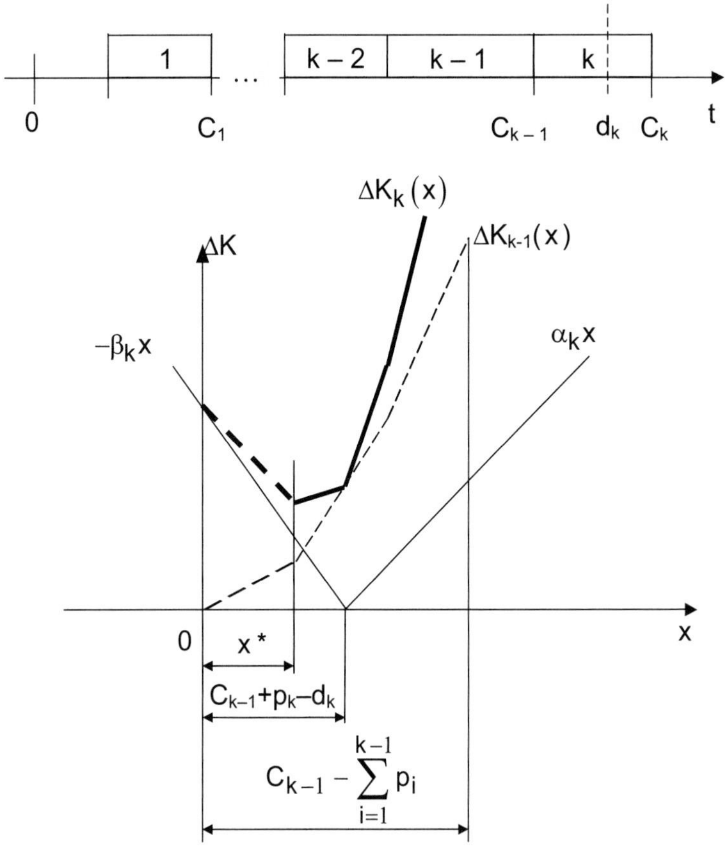

Fig. 4.5. Function $\Delta K_k(x)$.

2. Set $y = C^*_{k-1} + p_k - d_k$.
3. If $y > 0$ then update function $\Delta K_k(x)$ according to formula (4.13), set $C^*_k = C^*_{k-1} + p_k - x^*$, and move the origin to point $(x^*, K_k(x^*))$, else update function $\Delta K_k(x)$ according to formula (4.12) and set $C^*_k = d_k$.
4. If $k < n$ then set $k = k + 1$ and go to step 2.
5. Set $k = n - 1$ and $C_n = C^*_n$.
6. Set $C_k = \min\{C^*_k, C_{k+1} - p_{k+1}, C^*_{k+1} - p_{k+1}\}$

Algorithm 4.2 may be implemented to run in $O(nlogn)$ time using a special data structure (see [237]). Let us illustrate the algorithm with an example.

Example 4.3. Consider a sequence of $n = 4$ tasks with processing times $p_1 = 2$, $p_2 = 5$, $p_3 = 4$, $p_4 = 3$, due dates $d_1 = 5$, $d_2 = 13$, $d_3 = 15$, $d_4 = 17$, unit earliness costs $\alpha_1 = 2$, $\alpha_2 = 1$, $\alpha_3 = 3$, $\alpha_4 = 2$, and unit tardiness costs $\beta_1 = 1$, $\beta_2 = 1$, $\beta_3 = 2$, and $\beta_4 = 1$.

In the first iteration $y = 0 + 2 - 5 = -3 < 0$, so $C_1^* = d_1 = 5$, $\Delta K_1(x) = 2x$ for $0 \leq x \leq 3$. For $k = 2$ we obtain $y = 5 + 5 - 13 = -3 < 0$, hence $C_2^* = d_2 = 13$ and, according to formula (4.12),

$$\Delta K_2(x) = \begin{cases} x & \text{if } 0 \leq x \leq 3 \\ 3x & \text{if } 3 \leq x \leq 6. \end{cases} \tag{4.14}$$

For the third task we have $y = 13 + 4 - 15 = 2 > 0$ and

$$\Delta K_3(x) = \begin{cases} x + 2(2 - x) & \text{if } 0 \leq x \leq 2 \\ x + 3x & \text{if } 2 \leq x \leq 3 \\ 3x + 3x & \text{if } 3 \leq x \leq 6. \end{cases} \tag{4.15}$$

This function attains its minimum at $x^* = 2$, so $C_3^* = 15$. After moving the origin to point $(2, 2)$ we obtain:

$$\Delta K_3(x) = \begin{cases} 4x & \text{if } 0 \leq x \leq 1 \\ 6x & \text{if } 1 \leq x \leq 4. \end{cases} \tag{4.16}$$

Finally, for the last task we have $y = 15 + 3 - 17 = 1$ and

$$\Delta K_4(x) = \begin{cases} 4x + 1 - x & \text{if } 0 \leq x \leq 1 \\ 6x + 2x & \text{if } 1 \leq x \leq 4. \end{cases} \tag{4.17}$$

Function $\Delta K_4(x)$ attains its minimum at time $x^* = 0$, so $C_4^* = 18$. We move the origin to point $(0, 1)$. Now, we proceed to the second phase and assign the task completion times. We have $C_4^* = 18$, $C_3 = \min\{15, 18 - 3, 18 - 3\} = 15$, $C_2 = \min\{13, 15 - 4, 15 - 4\} = 11$, and $C_1 = \min\{5, 13 - 5, 11 - 5\} = 5$.

As we mentioned above, if the sequence is not given then the problem of minimizing function (4.1) is NP-hard. The properties of optimal schedules, proved for the problem with a common due date, do not hold in the case of task dependent due dates. Namely, idle time may occur in an optimal schedule and the sequence of tasks in an optimal schedule does not have to be V-shaped. The following properties are used as dominance rules in a branch and bound algorithm presented in [132].

Property 4.1 *If all tasks are tardy then an optimal sequence is obtained by ordering tasks according to non-decreasing values of* p_i/β_i, $i = 1, 2, \ldots, n$.

Property 4.2 *If all tasks are early then an optimal sequence is obtained by ordering tasks according to non-increasing values of* pi/α_i, $i = 1, 2, \ldots, n$.

In order to illustrate some difficulties emerging in the analysis of the earliness/tardiness problem with task dependent due dates and arbitrary weights, let us consider the following example.

Example 4.4. Consider a set of $n = 3$ tasks with processing times $p_1 = 3$, $p_2 = 4$, $p_3 = 4$, due dates $d_1 = 6$, $d_2 = 10$, $d_3 = 13$, unit earliness costs $\alpha_1 = 30$, $\alpha_2 = 7$, $\alpha_3 = 1000$, and unit tardiness costs $\beta_1 = 1$, $\beta_2 = 10$, $\beta_3 = 1000$. Let us consider two partial schedules consisting of tasks 1 and 2, presented in Figure 4.6. If task 1 precedes task 2, a schedule with zero cost and completion time equal to 10 is optimal. If task 2 precedes task 1, then an optimal schedule completes at time 13, and its cost equals 7. We may expect that the first schedule dominates the second one, in the sense that in any optimal schedule task 1 precedes task 2.

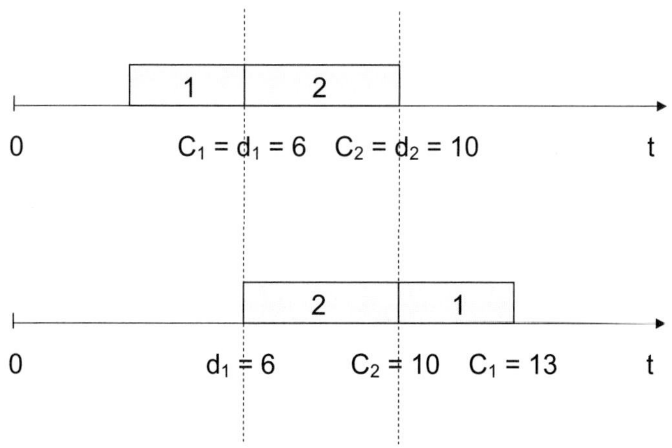

Fig. 4.6. Partial schedules of the sequences $\{1, 2\}$ and $\{2, 1\}$.

Let us now schedule task 3 as the last one, adding it to both schedules and finding optimal completion times according to Algorithm 4.2. The resulting schedules are presented in Figure 4.7. The sequence of tasks $\{1, 2, 3\}$ may be scheduled at the cost of minimum 37 units, while the sequence $\{2, 1, 3\}$ may be scheduled at the cost of 31 units.

Moreover, changing the weights may result in a different optimal sequence as shown in the following example.

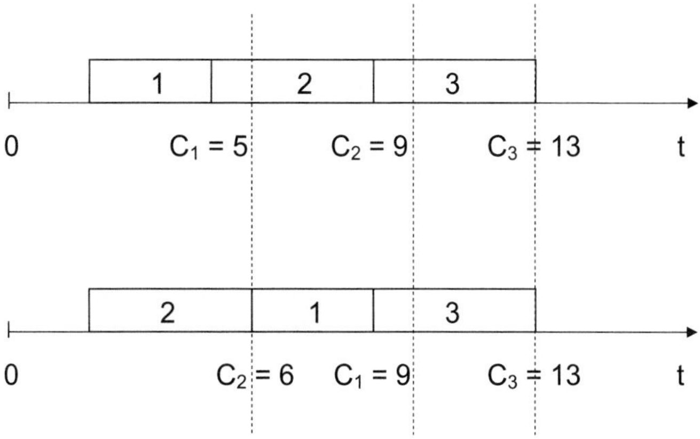

Fig. 4.7. Schedules of the sequences $\{1, 2, 3\}$ and $\{2, 1, 3\}$.

Example 4.5. Consider $n = 4$ tasks with processing times $p_i = 1$, $i = 1, 2, 3, 4$, due dates $d_1 = 1$, $d_2 = 2$, $d_3 = 2$, $d_4 = 2$, and unit earliness and tardiness costs $\alpha_1 = \beta_1 = 1$, $\alpha_i = \beta_i = 2$, $i = 2, 3, 4$. An optimal sequence is $1, 2, 3, 4$ and the cost of the schedule equals 6. However, if the unit earliness and tardiness costs increase to $\alpha_1 = \beta_1 = 2$, $\alpha_i = \beta_i = 8$, $i = 2, 3, 4$, then the cost of the schedule for sequence $1, 2, 3, 4$ equals 24, while the cost of the schedule for the sequence $4, 3, 2, 1$ equals 22.

Example 4.5 shows that even if the priorities of tasks do not change, only the relevant importance of a task decreases, an optimal sequence is no more optimal. This problem is investigated in more detail in [178, 179] and [212].

Concluding, we can see that there are not many dominance rules that can be used in solving the problem with task dependent due dates and arbitrary unit earliness and tardiness costs.

A dynamic programming algorithm to solve the problem with the objective to minimize function (4.1) is proposed in [131]. We present it briefly below. Let us first determine the scheduling horizon H. If it is not specified in the problem formulation, it can be calculated as $H = \max\{d_i\} + \sum_{i=1}^{n} p_i$. Let us assume that $n - k$ tasks are scheduled in the interval [0, s], $s \leq H$. We would like to schedule set J_k of the remaining k tasks optimally, starting not earlier than at time s. If task $j \in J$ is the task scheduled first in J and its processing starts at time $t \geq s$ then $C_j = t + p_j$, and the cost of scheduling task j equals $h_j(t) = \alpha_j \max\{0, d_j - t - p_j\} + \beta_j \max\{0, t + p_j - d_j\}$.

Let us assume further that $h^*(J_k \setminus \{j\}, s)$ is the optimal cost of scheduling set $J_k \setminus \{j\}$ starting not earlier than at time s. At each stage k, $k = 1, 2, \ldots, n$, i.e. with k tasks in set J_k, we have to choose task $j \in J_k$ to be scheduled first and its starting time ($t \geq s$), so that the value of the cost function is minimum. Therefore we find

$$h^*(J_k, s) = \min\{h_j(t) + h^*(J_k \setminus \{j\}, t + p_j) : j \in J, s \leq t \leq H\}. \quad (4.18)$$

Equation (4.18) is the recursive relation of the dynamic programming procedure. Initially, it is assumed that $h^*(\emptyset, s) = 0$, $s = 0, \ldots, H$. The optimal solution is found for $h^*(J_n, 0)$.

It is easy to observe that the number of decisions to be examined can be reduced by considering only feasible values of s. Namely, we know that the last task has to be completed by H, so it is sufficient to consider $s \leq H - \sum_{j \in J_k} p_j$, for each k. Nevertheless, at each stage k, $k = 1, 2, \ldots, n$, all k-element subsets of the set of tasks have to be examined, so the complexity of the algorithm is $O(n2^n H)$. For large n the proposed dynamic programming procedure can be more effective than the full enumeration, whose complexity is $O(n!)$. Moreover, it should be noticed that for many special cases, the time horizon H can be significantly reduced and Properties 4.1 and 4.2 can be applied in order to further decrease the computational effort of the algorithm.

Branch and bound algorithms solving the problem are proposed in [78], [131, 132] and in [219]. Davis and Kanet in [78] use a very simple lower bound based on the optimal solution of a partial schedule. Their algorithm solves instances with up to 8 tasks. In [132], where it is reported that instances with 20 tasks can be solved, tighter lower bounds are proposed.

Bülbül et al. in [37] consider the problem with arbitrary task ready times and propose lower bounds based on the following relaxation of the problem. They divide each task into unit-time tasks and associate weight with each task. Then they solve a corresponding transportation problem. The solution of the transportation problem is the lower bound for the scheduling problem under consideration. Computational experiments show that this approach is fast and effective.

Heuristic algorithms developed to solve the scheduling problem with the objective function (4.1) include neighborhood search proposed in [99], genetic algorithm developed in [169], tabu search algorithms proposed in [128, 129] and [133], and a local search algorithm examined in [183].

Lee and Choi in [169] propose a genetic algorithm for solving the scheduling problem with task dependent due dates. They provide an algorithm to find an optimal timing for a given sequence of tasks of complexity $O(n^2)$. The computational experiments show that the genetic algorithm finds optimal solutions for small instances and outperforms the heuristic by Yano and Kim presented in [252, 253].

James and Buchanan in [129] compare several tabu search implementations. The common concept is the compression of the solution space based on a binary representation of a solution. The binary vector indicates which tasks are early and which are tardy. Using this vector, a heuristic algorithm finds a permutation of tasks and the vector of completion times simultaneously. The algorithm solves instances with up to 250 tasks with the computational time bounded by 2 hours and 5 minutes for the largest instances.

Mazzini and Armentano in [183] propose a two stage heuristic. In the first stage a feasible solution is constructed, and in the second stage local search is applied to this solution. The computational experiments reported in [183] show that instances with 80 tasks may be solved in less than one second. The maximum deviation from optimal solutions found for small instances (with maximum 12 tasks) is 28% in the worst case. The maximum deviation from solutions found by the pairwise interchange heuristic proposed in [253] is 28% in the worst case. The pairwise interchange heuristic requires much more computational time, especially for large instances. The tabu search algorithm presented in [133] solves instances with up to 1000 tasks in 91 seconds, with optimal solutions found for all instances with up to 20 tasks.

Koulamas shows in [150] that the earliness/tardiness problem with due windows is NP-hard and propose a heuristic algorithm to find suboptimal solutions. Wan and Benjamin in [248] propose a tabu search algorithm for the problem of minimizing the earliness/tardiness cost on a single machine with due windows.

Coleman in [71] considers the single machine earliness/tardiness problem with sequence dependent setup times. The problem is formulated as a mixed integer program. Instances with no more than 8 tasks can be solved using the LINDO package.

Kanet and Sridharan in [143] investigate the problem of n tasks with arbitrary ready times and sequence dependent setup times to be scheduled on m parallel machines. The objective is to mimimize the total earliness and tardiness plus the setup costs. They propose a genetic algorithm which finds suboptimal solutions to the problem.

4.1.2 Proportional weights

Yano and Kim in [253] consider the problem with the objective to minimize function (4.1) with proportional weights, i.e. $\alpha_i = \alpha p_i$ and $\beta_i = \beta p_i$, $i = 1, \ldots, n$, $\alpha, \beta \geq 0$. Yano and Kim prove the following property that holds for the considered class of problems.

Property 4.3 *For adjacent tasks i and j, it is optimal to sequence task i before task j if*

$$\alpha_i/\alpha_j \leq p_i/p_j,$$
$$p_i/p_j \leq \beta_i/\beta_j,$$

and

$$\alpha_i p_j + \beta_j p_i \leq (\beta_j + \alpha_j)(C_j - C_i + p_i),$$

or

$$\alpha_i + \beta_i \leq \alpha_j + \beta_j.$$

Yano and Kim compare five heuristics to find an optimal sequence of tasks. The following sorting rules are considered:

- EDD (earliest due date) - tasks are ordered according to non-decreasing d_i,
- MDD (modified earliest due date),
- EST (earliest starting time) - tasks are ordered according to non-decreasing $d_i - p_i$,
- PREC (precedence) - tasks are ordered according to Property 4.3,
- INT (interchange) - pairwise interchanges are applied to the best sequence of EDD, MDD, EST and PREC.

The PREC rule is realized as follows. For all pairs of tasks (i, j) it is checked whether task i should precede task j according to Property 4.3. If the answer is affirmative, then the priority index of task i is increased by 1, and decreased by 1, otherwise. Finally, tasks are ordered according to non-increasing priorities. The procedure INT is applied to the pairs $(k, k+1)$, $k = 1, \ldots, n-1$, of consecutive tasks in the sequence. If an interchange improves the schedule it is accepted, and the next pair of tasks is checked. The optimal completion times are calculated for each sequence. The results of an extensive computational experiment reported in [253] show that the heuristic INT almost always finds solutions as good as those found by the branch and bound algorithm with one hour limit of the computational time.

The branch and bound algorithm proposed in [253] schedules tasks from set J, starting from the last one. Each node corresponds to a partial sequence J_k of k tasks. The optimal cost of scheduling this sequence can be found e.g. using Algorithm 4.2, under the constraint that the processing of the first task in J_k starts not earlier than the sum of processing times of tasks not yet sequenced, $\sum_{i \in J \setminus J_k} p_i$. It is the lower bound to the cost of the schedule of sequence J_k. The lower bound to the cost of scheduling set $J \setminus J_k$ of unscheduled tasks is calculated as the unavoidable earliness and tardiness that occurs because one or more tasks would overlap if they were scheduled with $C_i = d_i$. The lower bound is calculated as the sum of the lower bounds for the sequence J_k and for set $J \setminus J_k$. The algorithm solves all instances with up to 15 tasks and 50% of the instances with 20 tasks within the one hour limit of the computational time.

An efficient heuristic for the problem with proportional weights is proposed by Szwarc and Mukhopadhyay in [236]. The algorithm runs in two stages. In the first stage a sequence of single and multijob blocks is built. In the second stage the number of candidate sequences is reduced and optimal completion times of tasks in each sequence are found. Finally the best schedule among those sequences is selected . The solution procedure tested on a PC on instances with 40 and 50 tasks proves to be very fast.

4.1.3 Mean absolute lateness

Szwarc in [230], Fry et al. in [101, 102], Chang in [46], and Kim and Yano in [149] consider a special case of the scheduling problem with the objective to minimize function (4.1), i.e. one where $\alpha_i = \beta_i = 1$, $i = 1, \dots, n$. The objective function takes the form (4.19) and is called mean absolute lateness (MAL).

$$\frac{1}{n} \sum_{i=1}^{n} |C_i - d_i| \tag{4.19}$$

Fry et al. propose a heuristic procedure based on several properties of optimal schedules for the mean absolute lateness. The first two properties follow immediately from Properties 4.1 and 4.2 for identical unit earliness and tardiness costs.

Property 4.4 *Consider a sequence with two adjacent tasks i and j. If $p_i \geq p_j$, and i and j are early regardless of sequence, then there exists an optimal schedule in which task i precedes task j.*

Property 4.5 *Consider a sequence with two adjacent tasks i and j. If $p_i \leq p_j$, and i and j are tardy regardless of sequence, then there exists an optimal schedule in which task i precedes task j.*

In addition, the following properties hold.

Property 4.6 *Consider a sequence with two adjacent tasks i and j. If in the sequence in which task i precedes task j, task i is tardy and task j is early, then the reverse order never decreases the mean absolute lateness.*

Property 4.7 *Consider a sequence with two adjacent tasks i and j. If in the sequence in which task i precedes task j, task i is early, task j is tardy and $d_j - p_j \geq d_i - p_i$, then the reverse order never decreases the mean absolute lateness.*

Property 4.8 *Consider a sequence with two adjacent tasks i and j, such that i is tardy regardless of the sequence, and j is tardy only if i precedes j. If $p_i \leq p_j$ and $(d_j - \sum_{k \in prec(i)} p_k) \geq (p_j - p_i)/2$, where $prec(i)$ is the set of tasks preceeding task i in the sequence, then the absolute lateness with task i preceeding task j is never greater than with task j preceeding task i.*

The heuristic algorithm proposed in [101] enumerates a set of solutions obtained using three different *rules* to construct initial sequences and three different interchange *strategies*. The following *rules* are used to construct the initial sequence:

1. order tasks according to the non-decreasing due date;
2. order tasks according to the non-decreasing value of the start time calcutated as $d_i - p_i$;
3. order tasks randomly.

The following interchange *strategies* are examined in the computational experiment reported in [101]:

1. switching always begins in the first position in the sequence;
2. switching always begins in the last position in the sequence;
3. switching always chooses the most favourable pair.

Each strategy is applied to all three sequences. Optimal mean absolute lateness is calculated for each sequence and the best one is chosen.

Algorithm 4.6 (API [101]).

1. Set K at some large number. Set $i = 0$.
2. Set $i = i + 1$.
3. Find the initial sequence of tasks using *rule* i.
4. Set $k = 0$.
5. Set $k = k + 1$.

6. Choose the next pair of tasks to be interchanged using *strategy k*.
7. Calculate optimal mean absolute lateness K' of the current sequence using Algorithm 4.2.
8. If $K' < K$ then set $K = K'$.
9. If $k < 3$ then go to step 5.
10. If $i < 3$ then go to step 2.
11. Stop.

Solutions generated by the heuristic are compared with optimal solutions obtained by a branch and bound algorithm for instances with up to 16 tasks. In 122 instances out of 192 the solutions found by the heuristic are optimal.

Kim and Yano in [149] propose a branch and bound and heuristic algorithms to solve the MAL problem. They formulate some new properties of the problem, which they later use in their algorithms.

Property 4.9 *Consider two tasks, i and j, with* $d_i \leq d_j$. *If there is a conflict between these tasks when they are scheduled with* $C_i = d_i$ *and* $C_j = d_j$ *(that is* $d_j - d_i < p_j$*), then there exists an optimal schedule such that either* $C_i = d_i$ *or* $C_j = d_j$.

Property 4.10 *If there is a conflict between two tasks, i and j, when they are scheduled with* $C_i = d_i$ *and* $C_j = d_j$, *then the sum of earliness and tardiness is not less than the duration of the time conflict.*

Property 4.11 *Consider two tasks, i and j, with* $d_i \leq d_j$. *If there is a conflict between them when they are scheduled with* $C_i = d_i$ *and* $C_j = d_j$, *then*

(i) scheduling task i before task j is better if $p_j - p_i < 2(d_j - d_i)$, *and*
(ii) scheduling task i before task j is better if $p_j - p_i > 2(d_j - d_i)$.

Property 4.12 *If there are conflicts among two or more tasks when a set of tasks is scheduled with* $C_i = d_i$ *for* $i = 1, \ldots, n$, *then the sum of earliness and tardiness is greater than or equal to* $\sum_{k \geq 2} (k-1)t_k$, *where* t_k *is the length of time during which k tasks overlap.*

The way of calculating the lower bound resulting from ovelaps between tasks, given in Property 4.12, is illustrated in Figure 4.8.

The branch and bound algorithm proposed Kim and Yano in [149] uses branching rule similar to the one used in the algorithm proposed in [253]. Each node at level k of the branching tree represents a partial sequence J_k of the last k tasks. Not more than $n - k$ successor nodes may be created. Each successor node corresponds to a sequence of $k+1$

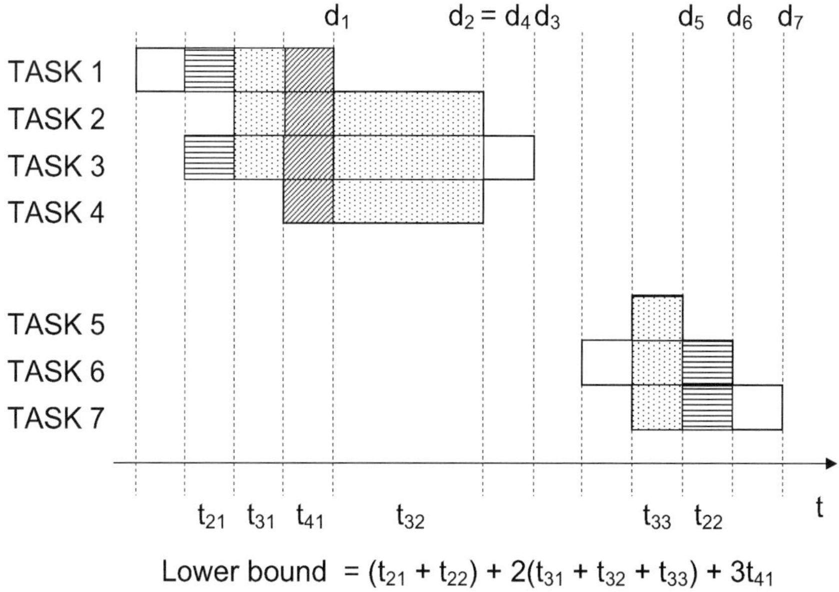

Fig. 4.8. Calculating the lower bound resulting from ovelaps between tasks.

tasks constructed by adding an unscheduled task at the beginning of sequence J_k. The depth-first branching rule is used with ties broken in favor of the node with the minimum lower bound. Nodes corresponding to dominated sequences are discarded. The initial upper bound may be obtained by any of the heuristics described below. Two schemata for calculating the lower bounds are used. One of them is tighter, but requires more computational time, and the other one is simpler to calculate, but it is less accurate. In both schemata the set of tasks is divided into three mutually disjoint and collectively exhaustive subsets, and the lower bound is the sum of lower bounds calculated independently for each subset. The three subsets are:

(a) the tasks in the partial sequence J_k;
(b) the set of unscheduled tasks with due dates larger than the starting time of the partial sequence J_k;
(c) all other tasks.

An optimal schedule of sequence J_k may be found in different ways, e.g. using Algorithm 4.2. However, the optimal starting time may not be feasible, since the tasks in set $J \setminus J_k$ must be completed before the first task in J_k starts. Let us denote by s the optimal feasible starting time of sequence J_k, i.e. maximum of the optimal starting time of sequence J_k and the total processing time of all unscheduled tasks.

In the first schema, the lower bound to the cost of the schedule of tasks in J_k is obtained as an optimal mean absolute lateness for the given sequence under the assumption that the first task in the sequence starts processing not earlier than at time s.

Let us now consider the set of tasks defined in (b). Assume that the schedule of sequence J_k starts at time $t \geq s$. We denote by $J^b(t)$ the set of tasks with due dates larger than t. The lower bound to the earliness cost is obtained by ordering the tasks in $J^b(t)$ according to the longest processing time with no idle time inserted and the last task completed at time t, and calculating the corresponding earliness cost. We add to the earliness cost of set $J^b(t)$, the cost of scheduling sequence J_k under the assumption that the first task in J_k starts at time t. This sum is calculated for each $t \geq s$ for which set $J^b(t)$ is nonempty. The first component of the lower bound is the minimum of the sum over t.

The second component, the lower bound to the earliness and tardiness cost of the remaining tasks is calculated according to Property 4.12 as the cost of the unavoidable overlaps.

Finally, the lower bound is the sum of the two components.

Let us illustrate the calculation of the lower bound with an example.

Example 4.7. Consider $n = 6$ tasks with processing times $p_1 = 1$, $p_1 = 1$, $p_2 = 1$, $p_3 = 1$, $p_4 = 1$, $p_5 = 2$, $p_6 = 1$, and due dates $d_1 = 1$, $d_2 = 1$, $d_3 = 6$, $d_4 = 8$, $d_5 = 10$, and $d_6 = 11$. Suppose that we consider the node of the branching tree at level 2, with $J_2 = \{3, 6\}$. The optimal start time of the partial schedule for J_2 is $t = 5$, and its cost equals 0. In this case the due dates of tasks 4 and 5 exceed t. These tasks are scheduled according to the longest processing time, and completed at $t = 5$. Their earliness cost equals 10. Finally, tasks 1 and 2 overlap during 1 time unit. We have to consider $5 \leq t \leq 10$, at which set $J^b(t)$ becomes empty. In Table 4.1 we present the calculation of the lower bound at this node.

In the second schema, the lower bound is calculated simply as

$$\sum_{i=1}^{N(t)} |d_i - t|. \tag{4.20}$$

where $N(t)$ is the number of tasks in J_k that are not early minus the number of early tasks in J_k that precede the first idle period in the optimal schedule of J_k which starts at t. This bound cannot be explained

Table 4.1. Calculation of the lower bound in Example 4.7

t	Cost of scheduling tasks in J_2	Cost of scheduling tasks in $J^b(t)$	Cost of scheduling the remaining tasks	Total
5	0	10	1	11
6	1	8	1	10
7	2	6	1	9
8	3	2	1	6*
9	4	1	1	6*
10	6	0	1	7

*) the lower bound equals 6

intuitively, but it can be proved (see [149]) that this bound is not greater that the lower bound calculated according to the first schema.

Computational experiments reported in [149] show that both lower bounds are almost equally effective. The branch and bound algorithm solves instances with up to 20 tasks in one hour.

The heuristic algorithms tested in a computational experiment reported in [149] include the heuristics EDD, EST, PREC, and INT proposed in [253], API proposed in [101], and the following ones:

- NEH - an algorithm proposed by Nawaz et al., presented below;
- PI - pairwise interchange - similar to API, but instead of checking the interchanges of adjacent tasks only, it checks interchanges of all pairs of tasks;
- TS - tabu search algorithm.

Algorithm 4.8 (NEH [195]).

1. Calculate the optimal mean absolute lateness K for initial sequence S.
2. Set $k = 1$.
3. Set $l = 1$.
4. Create new sequence S' with task k inserted in position l in S. Calculate the optimal mean absolute lateness K' for sequence S'.
5. If $K' < K$ then set $K = K'$ and $S = S'$.
6. If $l < n$ then set $l = l + 1$ and go to step 4.
7. If $k < n$ then set $k = k + 1$ and go to step 3, else stop.

The initial sequence may influence the result of the NEH algorithm. In [149] the following initial sequences are examined: EDD (earliest due date), LDD (latest due date), SPT (shortest processing time), and LPT (longest processing time). The computational experiment shows that simple sorting heuristics may produce solutions that differ even by about 30% from optimum, but the pairwise interchange improves the results significantly (to less than 1% deviation from optimum). Tabu search requires more computational time than pairwise interchange, but gives slightly better results.

Chang in [46] proposed tighter lower bounds calculated from the overlaps of tasks. The branch and bound algorithm using these bounds solves instances with up to 45 tasks in 6000 seconds on a VAX/6510. The set of test instances is generated using the method proposed by Potts and Van Wassenhove in [207].

Ventura and Radhakrishnan in [243] formulate the MAL problem as a 0-1 linear integer program and propose a heuristic algorithm to solve it, based on Lagrangian relaxation. The computational experiments show that the heuristic finds near optimal solutions, but computational times are larger than the tabu search algorithm proposed in [149].

Schaller in [213] considers the objective function formulated as the sum of the earliness plus the squared tardiness of tasks. He proposes a branch and bound algorithm and a heuristic to solve the problem.

4.1.4 Maximizing the number of just-in-time tasks

Single machine

The cost of the schedule considered so far in this Chapter depended on the total earliness and tardiness of tasks. In this section we mention another objective related to the earliness/tardiness scheduling which is maximization of the number of tasks completed on time.

Yoo and Martin-Vega in [257] consider the problem of minimizing the number of tasks completed outside a given due date window, with idle time allowed. Let $[\underline{d}_i, \overline{d}_i]$ be the due window of task i. The cost of scheduling task i is determined as follows.

$$U_i = \begin{cases} 0 & \text{if } \underline{d}_i \leq C_i \leq \overline{d}_i \\ 1 & \text{otherwise} \end{cases} \tag{4.21}$$

The objective is to minimize function (4.22).

$$f(S) = \sum_{i=1}^{n} U_i \tag{4.22}$$

The problem was proved to be NP-hard by Hiraishi et al. in [122]. Yaoo and Vega propose the following heuristic algorithm to solve this problem with arbitrary release dates r_i. The algorithm schedules the tasks backwards, eliminating tasks which cannot be scheduled on time and then constructs the final schedule by reversing the order of tasks scheduled on time and appending it with tardy tasks.

Algorithm 4.9 (Yoo and Martin-Vega [257]).

1. Obtain a feasible sequence of tasks by scheduling a task with the earliest due date, whenever the machine becomes available.
2. Find the first task j which is not completed in its due window. If there is no such task then go to step 5.
3. Reschedule tasks $1, \ldots, j$, as to obtain a partial schedule with total cost equal to zero. If such a partial schedule does not exist then go to step 4, else consider the new sequence as the current sequence and go to step 2.
4. Find tasks in sequence $1, \ldots, j$, whose removal results in a partial schedule with no more tasks completed outside the due window. Remove the task whose removal results in the shortest completion time of the partial schedule.
5. Construct the final sequence by reversing the order of the current sequence and adding all removed tasks at the end of the sequence.

Computational experiments reported in [257] show that the algorithm finds solutions much better than a simple heuristics based on the EDD rule. No comparison with exact solutions is reported.

Parallel machines

Čepek and Sung in [44] consider the problem to maximize the number of tasks completed exactly at their due dates in the system of identical parallel machines. They develop a polynomial time algorithm to solve the problem. The problem is formulated as follows. Consider n non-preemptive, independent tasks and m identical parallel machines. Let us denote by p_i the processing time and by d_i the due date of task i, respectively. The cost of scheduling task i is determined as follows.

$$U_i = \begin{cases} 0 & \text{if } C_i = d_i \\ 1 & \text{otherwise} \end{cases} \tag{4.23}$$

The objective is to minimize function (4.24).

$$f(S) = \sum_{i=1}^{n} U_i \tag{4.24}$$

Algorithm 4.10 finds the optimal solution to this problem in $O(n^2)$ time. We denote by T the set of tardy tasks.

Algorithm 4.10 (Čepek and Sung [44]).

1. Index tasks according to non-decreasing order of values $d_i - p_i$. Set $k = 1$ and $T = \emptyset$.
2. If there is at least one machine available at time $d_k - p_k$ then schedule task k on any available machine, so that it completes exactly at time d_k and go to step 4.
3. Consider the set J_k of tasks occupying machines $\{1, \ldots, m\}$ at time $d_k - p_k$ and select the task $out(k)$, such that $d_{out(k)} = \max\{d_i : i \in J_k\}$. Replace task $out(k)$ with task k and schedule task k so that it completes at d_k. Add task $out(k)$ to the set of tardy tasks T.
4. If $k < n$ then set $k = k + 1$ and go to step 2.
5. Schedule tasks from set T on arbitrary machine when it becomes available.

Sung and Vlach in [229] consider a more general problem of maximizing the weighted number of tasks completed at the due date (4.25) in a system of unrelated parallel machines .

$$f(S) = \sum_{i=1}^{n} w_i U_i \tag{4.25}$$

An algorithm to solve this problem is proposed in [229]. The complexity of this algorithm is polynomially bounded as a function of the number of tasks, but exponential as a function of the number of machines.

4.1.5 Minimizing the maximum earliness/tardiness cost

Earlier in this book we considered various objective functions, which included the total earliness/tardiness cost. In some situations it may be more appropriate to minimize the maximum earliness/tardiness cost over all tasks, i.e. the objective function given by (4.26).

$$\max\{f^e(\max_i\{e_i\}), f^t(\max_i\{t_i\})\} \tag{4.26}$$

Moreover, from the point of view of practical applications it is interesting to consider more flexible constraints where no additional penalty is incurred if a task completes in a given interval. Sidney in [218] examines the problem of minimizing the objective function (4.26) with due windows. The problem is formulated as follows. If task i starts before a given target start time r_i then the earliness cost is incurred, if it is completed after the due date d_i, then the tardiness cost is incurred,

$i = 1, \dots, n$. Accordingly, the earliness and tardiness are defined as follows:

$$e_i = \max\{0, r_i - C_i - p_i\}, i = 1, \dots, n \tag{4.27}$$

$$t_i = \max\{0, C_i - d_i\}, i = 1, \dots, n \tag{4.28}$$

A special case where the time window has the property that if $r_i < r_j$ then $d_i \le d_j$, is solvable in polynomial time. The algorithm, of complexity $O(n^2)$, proposed by Sidney in [218] is presented below.

Algorithm 4.11 (Sidney [218]).

1. Order tasks so that $r_1 \le r_2 \le \dots \le r_n$ and $d_1 \le d_2 \le \dots \le d_n$.
2. Calculate e^*, t^* as a solution of the following set of equations:

$$\begin{cases} e^* + t^* = \max\limits_{i<j} \left\{ \sum\limits_{k=i}^{j} p_k - (d_j - r_i) \right\} \\ \\ f^e(e^*) = f^t(t^*) \end{cases} \tag{4.29}$$

3. $C_1 := r_1 + p_1 - e^*$;
4. Set $C_i = \max\{C_{i-1} + p_i, r_i + p_i - e^*\}$, $i = 2, \dots, n$.

Let us illustrate the algorithm with an example.

Example 4.12. Consider $n = 7$ tasks with processing times and due intervals presented in Table 4.2, and cost functions $f^e(x) = 2x$, $f^t(x) = 3x$, $x \ge 0$.

Table 4.2. Characteristics of tasks from Example 4.12

i	1	2	3	4	5	6	7
p_i	2	3	3	5	1	2	3
r_i	1	3	4	6	7	10	12
d_i	4	7	9	11	12	20	25

In Table 4.3 we present the values $\max\limits_{i<j} \left\{ \sum\limits_{k=i}^{j} p_k - (d_j - r_i) \right\}$ calculated for Example 4.12.

Table 4.3. Characteristics of tasks from Example 4.12

i \ j	1	2	3	4	5	6	7
1	-1	-1	0	3	3	-3	-5
2	-	-1	0	3	3	-3	-5
3	-	-	-2	1	1	-5	-7
4	-	-	-	0	0	-6	-8
5	-	-	-	-	-4	-10	-12
6	-	-	-	-	-	-8	-10
7	-	-	-	-	-	-	-10

The maximum value in Table 4.2 equals 3, so the solution of equations (4.29) is $e^* = 1.8$ and $t^* = 1.2$. Consequently, $C_1 = 1 + 2 - 1.8 = 1.2$, $C_2 = 4.2$, $C_3 = 7.2$, $C_4 = 12.2$, $C_5 = 13.2$, $C_6 = 15.2$, and $C_7 = 18.2$. It is easy to see that the first task starts before time zero. Since usually such schedule is not feasible, we obtain a feasible schedule by delaying all tasks by 1.8.

4.1.6 Scheduling with additional resources

In this section we consider the situation where processing time of a task depends on the amount of an additional resource allotted to the task. Cheng and Janiak in [61] examine the following problem with arbitrary ready times and resource dependent processing times. Let us consider n nonpreemptive and independent tasks. Let r_i denote the ready time of task i, u_i the amount of the resource allotted to task i, and γ the resource unit cost. Processing time of task i, $i = 1, \ldots, n$, is defined as $p_i = b_i - a_i u_i$. The available amount of resource u_i is limited, i.e. $\underline{u}_i \leq u_i \leq \overline{u}_i$. The objective function is defined as the total cost of earliness, tardiness and resource consumption (4.30).

$$\sum_{i=1}^{n} (\alpha_i e_i + \beta_i t_i + \gamma u_i) \tag{4.30}$$

Notice that for $\gamma = 0$ and $\alpha_i = 0$, and $a_i = 0$, $i = 1, \ldots, n$, the problem reduces to the single machine weighted tardiness problem which is known to be NP-hard (see [167]).

A heuristic algorithm is proposed in [61] to minimize function (4.30) with resource dependent processing times. The algorithm first uses some greedy procedure to find a feasible schedule and then tries to reduce the schedule cost by applying one of the six improvement operations:

1. shift the task to the left (processing time remains unchanged);
2. shift the task to the right (processing time remains unchanged);
3. start the task later (processing time is reduced, so that the completion time remains unchanged);
4. start the task earlier (processing time is extended, so that the completion time remains unchanged);
5. complete the task later (processing time is extended, so that the start time remains unchanged);
6. complete the task earlier (processing time is reduced, so that the start time remains unchanged).

The change of the objective function, η_i^j, $j = 1, \ldots, 6$, $i = 1, \ldots, n$, resulting from any of the above operations may be recursively calculated for each task in a given sequence. Below we show, for example, how to calculate the values of η_i^1 (shifting task i to the left) and η_i^6 (completing task i earlier without changing its starting time), $i = 1, \ldots, n$. We assume that tasks are indexed according to their positions in the sequence. The coefficients η_i^1, $i = 1 \ldots, n$ are calculated as follows:

$$\eta_i^1 = \begin{cases} \infty & \text{if } C_i = r_i + p_i \\ \min\{\eta_{i-1}^1, \eta_{i-2}^6\} - \beta_i & \text{if } C_i > r_i + p_i, C_i = C_{i-1} + p_i, C_i > d_i \\ \min\{\eta_{i-1}^1, \eta_{i-2}^6\} + \alpha_i & \text{if } C_i > r_i + p_i, C_i = C_{i-1} + p_i, C_i \le d_i \\ -\beta_i & \text{if } C_i > r_i + p_i, C_i > C_{i-1} + p_i, C_i > d_i \\ \alpha_i & \text{if } C_i > r_i + p_i, C_i > C_{i-1} + p_i, C_i \le d_i \end{cases} \tag{4.31}$$

The recursion is initiated with

$$\eta_1^1 = \begin{cases} \infty & \text{if } C_1 = r_1 + p_1 \\ -\beta_1 & \text{if } C_1 > r_1 + p_1, C_1 > d_1 \\ \alpha_1 & \text{if } C_1 > r_1 + p_1, C_1 \le d_1 \end{cases} \tag{4.32}$$

The coefficients η_i^6, $i = 1 \ldots, n$ are calculated as follows:

$$\eta_1^6 = \begin{cases} \infty & \text{if} \quad u_i = \overline{u}_i \\ 1/a_i - \beta_i & \text{if} \quad u_i < \overline{u}_i, C_1 > d_1 \\ 1/a_i + \alpha_i & \text{if} \quad u_i < \overline{u}_i, C_1 \leq d_1 \end{cases} \tag{4.33}$$

Using the same reasoning as above, it is easy to calculate all the remaining coefficients. The algorithm is presented below.

Algorithm 4.13 (Cheng and Janiak [61]).

1. Construct an initial sequence of tasks, S.
2. Find completion time and resource allocation for the first task from sequence S, such that the cost of scheduling this task is minimum. Set $k = 1$.
3. Assume that the completion times and resource allocations (i.e. also processing times) of tasks $1, \ldots, k$, are given by the current partial schedule. Set $l = 1$.
4. If scheduling task $k+1$ in position l violates neither start nor completion times of the scheduled tasks then calculate the minimum cost of such schedule K_l.
5. If $l < k+1$ then set $l = l+1$ and go to step 4, else set the current schedule to be the $(k+1)$-element partial schedule with minimum cost K_l.
6. If $k < n$ then set $k = k+1$ and go to step 3.
7. For each task i, $i = 1, \ldots, n$, calculate the coefficients η_i^j, $j = 1, \ldots, 6$.
8. Select the parameter $\eta_{i'}^{j'}$ with the minimum value. If this value is negative then go to step 9, else stop.
9. Execute the improvement operation defined by the coefficient $\eta_{i'}^{j'}$ and go to step 8.

The execution of the improvement operation may affect other tasks. Thus, the set J_{IMP} of all tasks affected by this operation is constructed. Further, the admissible length of the operation (shift, reduction or extension of a task) is found for each task in set J_{IMP}. The admissible length of operation j on task i is the maximum length that does not change the value of the relevant coefficient η_i^j. The minimum value over the admissible changes of all tasks is chosen as the executable length of the operation.

Seven ways of finding the initial sequence were tested in a computational experiment reported in [61]:

- according to non-increasing values of α_i;
- according to non-increasing values of β_i;
- according to non-increasing values of $\alpha_i + \beta_i$;
- according to non-increasing values of $\max\{\alpha_i, \beta_i\}$;
- according to non-decreasing values of d_i/α_i;
- according to non-decreasing values of r_i;
- according to non-increasing values of $b_i - a_i\overline{u}_i$;

The results of the computational experiment show that the quality of the solutions found by the algorithm depends on the initial sequence, although the differences are not significant. The difference between minimum and maximum relative deviation from the lower bound is big for all ways of obtaining the initial sequence. Average computational times are small and do not exceed one minute for instances with $n = 100$ tasks on an IBM PC AT.

4.1.7 Other models

Many different applications of the earliness/tardiness scheduling problems result in new models and algorithms. Some interesting formulations are presented in this section.

Minimizing total earliness with no tardy tasks

Chand and Schneeberger in [45] consider a problem of minimizing the total earliness with the constraint that no task is tardy. Let n be the number of tasks, and p_i, d_i, and α_i be the processing time, due date and unit earliness cost of task i, respectively. The problem is formulated as follows.

$$\begin{aligned} \text{minimize} \quad & \sum_{i=1}^{n} \alpha_i(d_i - C_i) \\ \text{subject to} \quad & d_i - C_i \le 0 \;\; i = 1, \ldots, n \\ & C_i - C_{i-1} \ge p_i \;\; i = 2, \ldots, n \\ & C_1 \ge p_1 \end{aligned}$$

The NP-hardness proof as well as two algorithms proposed to solve this problem are presented in [45]. One of the algorithms is a pseudopolynomial time dynamic programming procedure. The second one is a heuristic algorithm. The heuristic algorithm starts scheduling tasks from the last position. Let J be the set of unscheduled tasks.

Algorithm 4.14 (Chand and Schneeberger [45]).

1. Set $J = 1, \ldots, n$, $H = \max_i\{d_i\}$, and $k = n$.
2. Find task j such that $p_j/\alpha_j = \min_{i \in J, d_i \geq H}\{p_i/\alpha_i\}$.
3. Set $C_j = H$, $J = J \setminus \{j\}$, $k = k - 1$ and $H = \min\{C_j - p_j, \max_{i \in J}\{d_i\}\}$.
4. If $k = 0$ then stop, else go to step 2.

Chand and Schneeberger in [45] prove that the heuristic finds optimal solutions for the following special cases of the considered problem:

- $p_1/\alpha_1 = p_2/\alpha_2 = \ldots = p_i/\alpha_i$;
- for any pair of tasks if $p_i/\alpha_i \geq p_j/\alpha_j$ then $d_i \leq d_j$;
- $d_1 = d_2 = \ldots = d_n$.

The computational experiments reported in [45] show that the heuristic may find solutions far from optimal ones in the general case. However, if the due dates are relatively uniformly distributed over a wide range then the heuristic performs well.

Minimizing the total earliness and completion time

Fry et al. in [100] consider the problem of scheduling tasks, all ready for processing at time zero, with the objective function given by formula (4.34).

$$\sum_{i=1}^{n} (\alpha e_i + \gamma C_i) \tag{4.34}$$

where γ, is the unit cost of task completion time. Notice that the tardiness is not explicitly penalized, although large completion times are discouraged. The authors formulate the problem as a mixed integer program. Since the computational time required to solve the MIP problem grows almost exponentially with the number of tasks, exact solutions were found only for instances with for up to 15 tasks.

Another problem is examined by Keyser and Sarper in [148]. They consider the problem of scheduling n nonpreemptive and independent tasks with processing times p_i, due dates d_i and release dates r_i, $i = 1, \ldots, n$ on a single machine, minimizing the total cost incurred by task earliness e_i, tardiness t_i, and waiting time q_i. Earliness and tardiness are defined by formulas (2.6) and (2.7), respectively, and waiting time is defined by formula (4.35).

$$q_i = \begin{cases} C_i - p_i - r_i & \text{if } C_i - p_i \geq r_i \\ 0 & \text{otherwise} \end{cases} \tag{4.35}$$

The unit waiting cost is denoted by γ_i, $i = 1, \ldots, n$. In [148] the problem is formulated as the following mixed integer program.

$$\text{minimize} \sum_{i=1}^{n} (\alpha_i e_i + \beta_i t_i + \gamma_i q_i) \tag{4.36}$$

subject to

$$q_i - t_i + e_i = d_i - p_i - r_i, \quad i = 1, \ldots, n \tag{4.37}$$

$$C_i - q_i = p_i - r_i, \quad i = 1, \ \ldots, n \tag{4.38}$$

$$C_i - C_j + M(1 - y_{ij}) \geq p_i, \quad i = 1, \ldots, n, j = i + 1, \ldots, n \tag{4.39}$$

$$C_i - C_j + M(y_{ij}) \geq p_j, \quad i = 1, \ldots, n, j = i + 1, \ldots, n \tag{4.40}$$

where M is a large positive number and y_{ij} is a binary decision variable such that

$$y_{ij} = \begin{cases} 1 & \text{if task } i \text{ precedes task } j \text{ in the sequence,} \\ 0 & \text{otherwise.} \end{cases} \tag{4.41}$$

A simple heuristic to solve the scheduling problem with the objective formulated as (4.36) is proposed in [148]. The algorithm constructs three sequences, obtained by ordering the tasks according to (i) non-decreasing release times, (ii) non-increasing priorities o_i calculated from formula (4.42), and (iii) non-decreasing slack, calculated as $d_i - p_i - r_i$. In formula (4.42) t denotes the current time unit in the schedule, k is a look ahead factor, and $\overline{p}$ is the average task processing time. The priority rule (4.42) is adapted from the priority function given by Ow and Morton in [203].

$$o_i = \begin{cases} \dfrac{\alpha_i+\beta_i}{p_i+r_i} exp\left(\dfrac{\alpha_i+\beta_i+\gamma_i}{\alpha_i}\dfrac{\max\{0, d_i-t-p_i-r_i\}}{\overline{p}}\right), \\ \qquad \text{if } d_i-t-p_i-r_i \le \dfrac{(\alpha_i+\beta_i)k\overline{p}}{\alpha_i+\beta_i+\gamma_i}, \\ \dfrac{\alpha_i}{p_i+r_i}\left(\dfrac{\gamma_i+\beta_i}{\alpha_i} - \dfrac{\alpha_i+\beta_i+\gamma_i}{\alpha_i}\dfrac{\min\{k\overline{p}, d_i-t-p_i-r_i\}}{k\overline{p}}\right)^3, \\ \qquad \text{otherwise.} \end{cases} \tag{4.42}$$

Let us assume that tasks are indexed according to their order in the sequence. Two schedules are obtained from each sequence. In the first schedule the processing of task i starts as soon as possible, i.e. at time $\max\{r_i, C_{i-1}\}$. In the second schedule the processing of task i starts as late as possible, i.e. at time $\max\{d_i - p_i, r_i, C_{i-1}\}$, $i = 1, \ldots, n$. The adjacent pair interchange (Algorithm 4.6) is applied to the best sequence. The interchange starts from the beginning of the sequence. The solutions obtained by the heuristic are compared with optimal solutions found by solving the mixed integer program (4.36) - (4.40) using the LINDO package. Optimal solutions can be found only for instances with up to 6 tasks. For majority of instances (41 out of 48) the solutions generated by the heuristic are optimal.

Scheduling with machine vacations

One more variant of the problem is proposed by Mannur and Addagatla in [182]. Namely, they consider an additional constraint, called machine vacations, meaning that the machine is unavailable in given periods of time. Such machine vacations may be practically justified, e.g. by maintenance requests. Let us assume that there are m such periods of machine vacations, each defined by an interval $[t_k^l, t_k^r]$, $k = 1, \ldots, m$. The problem is formulated as follows:

$$\text{minimize} \quad \frac{1}{n}\sum_{i=1}^{n} |C_i - d_i| \tag{4.43}$$

subject to

$$C_i - C_j \ge p_i \text{ or } C_j - C_i \ge p_j, \quad i \ne j, i = 1, \ldots, n, j = 1, \ldots, n, \tag{4.44}$$

$$C_i \le t_k^l \text{ or } C_i \ge t_k^l + t_k^r, \quad i = 1, \ldots, n, k = 1, \ldots, m, \tag{4.45}$$

$$C_i \geq r_i + p_i, \quad i = 1, \ldots, n, k = 1, \tag{4.46}$$

$$C_i \geq 0, \quad i = 1, \ldots, n, k = 1. \tag{4.47}$$

Constraint (4.44) assures that only one task is processed at a time, constraint (4.45) prevents scheduling tasks during the machine vacations, and (4.46) guarantees respecting the task ready times. Constraint (4.47) assures that completion times C_i are non-negative. Mannur and Addagatla in [182] propose two heuristic algorithms and compare their effectiveness on the basis of a computational experiment.

In the first heuristic the schedule is constructed in consecutive time windows between machine vacation periods. Tasks are assigned to consecutive windows on the basis of their release times and processing times. As many tasks as possible are scheduled in each window, and the unscheduled tasks are added to the set of tasks assigned to the next window.

The second heuristic schedules tasks according to non-decreasing due dates. If a task cannot complete before the next vacation, it is moved to the next window. If the due date of a task falls in a vacation period then the task is scheduled either in the preceding or in the succeeding time window, depending on the incurred penalty. The first heuristic generates significantly better schedules than the second one, although requires longer computational time, especially for larger instances. Both heuristics solve instances with 150 tasks in milliseconds.

Scheduling with batch setup times

The problem of scheduling tasks in batches with setup times in order to minimize the total earliness/tardiness costs is considered by Chen in [48]. He assumes that all tasks in a batch have the same due date and there is a machine setup required between batches. The problem is formulated as follows. Lets us consider $b \geq 1$ batches of nonpreemptive and independent task to be processed on a single machine. Each batch consists of n_j tasks and has a due date d_j, $j = 1, \ldots, b$. A setup time s_{ij} is required whenever a task from batch j is processed immediately after a task from batch l, $j = 1, \ldots, b$, $l = 1, \ldots, b$. Let us denote by p_{ij} the processing time, and by C_{ij} the completion time of task i in batch j. We define the earliness and tardiness of task i in batch j as follows:

$$e_{ij} = \max\{0, d_i - C_{ij}\}, \tag{4.48}$$

$$t_{ij} = \max\{0, C_{ij} - d_i\}. \tag{4.49}$$

The unit earliness cost α_j and tardiness cost β_j are equal for all tasks in batch j, $j = 1, \ldots, n$. The problem is to find a schedule with minimum total earliness/tardiness cost given by (4.50).

$$\sum_{j=1}^{b}\sum_{i=1}^{n_j}(\alpha_j e_{ij} + \beta_j t_{ij}) \tag{4.50}$$

This problem is NP-hard even for two batches and unrestrictive due dates. NP-hardness is proved in [48] by reduction from the even-odd partition problem.

Several properties are proved in [48] for the special case of the problem with $d_j = d \geq \sum_{j=1}^{b}\sum_{i=1}^{n} p_{ij} + \max\{s_{ij}\}\sum_{j=1}^{b} n_j$, $j = 1, \ldots, n$.

Property 4.13 *If all batches have the same unrestrictive due date then the optimal schedule has the following properties:*

1. *there is no idle time in the schedule (excluding setup time);*
2. *each batch of tasks forms a V-shape;*
3. *some task completes at time d.*

A dynamic programming algorithm of complexity $O(n^{2b}/b^{2b-3})$ for solving this special case is given in [48]. A similar problem with controllable due dates is presented in Section 4.3.3.

4.2 Schedules without idle time

As we mentioned at the beginning of this chapter, there are situations in which idle time is not allowed. In this section we present the properties of optimal schedules and solution algorithms for the single machine scheduling problem with task dependent due dates and no idle time. In Section 4.2.1 we consider the general case with arbitrary unit earliness and tardiness costs. Section 4.2.2 deals with a special case of the scheduling problem with task independent unit costs.

4.2.1 Arbitrary weights

Abdul-Razaq and Potts in [1] introduce the following problem. Consider n independent and nonpreemptive tasks with arbitrary integer processing times p_i, $i = 1, \ldots, n$, to be scheduled on a single machine. Completing task i at time t incurs cost $f_i(t)$. Machine idle time is not permitted. The objective is to minimize the total cost calculated as follows:

$$\sum_{i=1}^{n} f_i(C_i) \tag{4.51}$$

Notice that if $f_i(t)$ is non-decreasing then the problem of minimizing function (4.51) is equivalent to the mean (or total) tardiness problem. Abdul-Razaq and Potts propose a dynamic programming algorithm for an arbitrary function $f_i(t)$, and perform computational experiments for $f_i(t)$ given by formula (4.52).

$$f_i(t) = \alpha_i \max\{0, d_i - t\} + \beta_i \max\{0, t - d_i\} \tag{4.52}$$

The dynamic programming algorithm is formulated as follows. Let us denote by $h^*(J_k)$ the minimum total cost of scheduling subset J_k of k tasks in the first k positions in the sequence. For each k-element subset of J_n the decision is made which task should be scheduled in position $k+1$. The recurrence relation is given by formula (4.53).

$$h^*(J_k) = \min_{i \in J_k} \left\{ h^*\left(J_k \setminus \{i\}\right) + f_i\left(\sum_{j \in J_k} p_j\right)\right\} \tag{4.53}$$

The initial condition is $h^*(\emptyset) = 0$, and the optimal solution is obtained for $h^*(J_n)$. The problem is equivalent to the problem of finding the shortest path in a graph in which nodes correspond to the subsets of tasks, and the length of the arc from node $J_k \setminus \{i\}$ to J_k equals $f_i(\sum_{j \in J_k} p_j)$. Obviously, the computational complexity of the algorithm is $O(2^n)$. Thus, only small instances may be solved.

Abdul-Razaq and Potts propose to derive from (4.52) a lower bound for the branch and bound algorithm. The idea is to relax the state space by mapping the states representing subsets of tasks onto the states representing the total processing time of tasks in the subset. We may regard $g^*(t)$ as the minimum cost of scheduling tasks in the interval $[0, t]$. The relaxed problem is solved by computing $g^*(\sum_{i=1}^{n} p_i)$ from the recurrence relation (4.54).

$$g_0^*(t) = \min_{i \in J_n} \{g_0^*\left(t - p_i\right) + f_i\left(t\right)\} \tag{4.54}$$

The initial condition is $g_0^*(t) = \infty$ for $t < 0$, and $g_0^*(0) = 0$. The complexity of the relaxed problem is $O(n \sum_{i=1}^{n} p_i)$. This problem is equivalent to the problem of finding the shortest path in a graph in which nodes correspond to time units $1, \ldots, \sum_{i=1}^{n} p_i$, and in which for each task there is an arc from node $t-p_i$ to node t, $t = p_i, \ldots, \sum_{j=1}^{n} p_j$ of length $f_i(t)$. The arc corresponds to scheduling task i in the interval $[t-p_i, t]$. In order to illustrate the relaxation let us consider the following example.

Example 4.15. Consider the problem of scheduling $n = 3$ tasks with processing times $p_1 = 2$, $p_2 = 3$, and $p_3 = 4$, due dates $d_1 = 4$, $d_2 = 5$, and $d_3 = 6$, earliness penalties $\alpha_1 = 1$, $\alpha_2 = 3$, and $\alpha_3 = 7$, and tardiness penalties $\beta_1 = 2$, $\beta_2 = 1$, and $\beta_3 = 8$. The graph corresponding to problem (4.52) is presented in Figure 4.9. The solutions of the recurrence equations are $f^*(\{1\}) = 2$, $f^*(\{2\}) = 6$, $f^*(\{3\}) = 14$, $f^*(\{1,2\}) = 2$, $f^*(\{1,3\}) = 3$, $f^*(\{2,3\}) = 14$, and $f^*(\{1,2,3\}) = 6$. Thus, the optimal sequence is $(1,3,2)$, and the cost of the optimal schedule is 6. The graph corresponding to problem (4.54) is presented in Figure 4.10. The shortest path includes nodes $(0,2,5,8)$ and is of length 5. The corresponding sequence of tasks is $(1,2,2)$. Obviously, it is not a feasible solution of the scheduling problem, since task 2 appears twice, while task 3 does not appear at all.

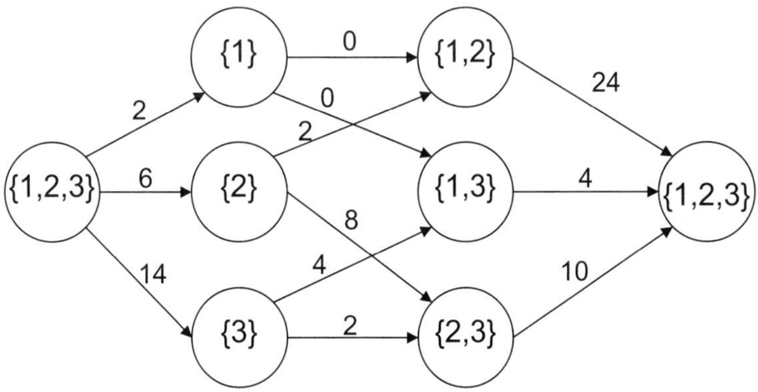

Fig. 4.9. Scheduling the kth task in case (i).

Abdul-Razaq and Potts prove in [1] that the solution of the relaxed problem provides a lower bound to the solution of the original problem. The further propose some procedures to improve the lower bound.

The first modification allows to avoid sequences in which the same task appears in adjacent positions.

The second modification aims to avoid scheduling a task more than once by adding a penalty λ_i to each scheduled task. The modified recurrence relation is the following.

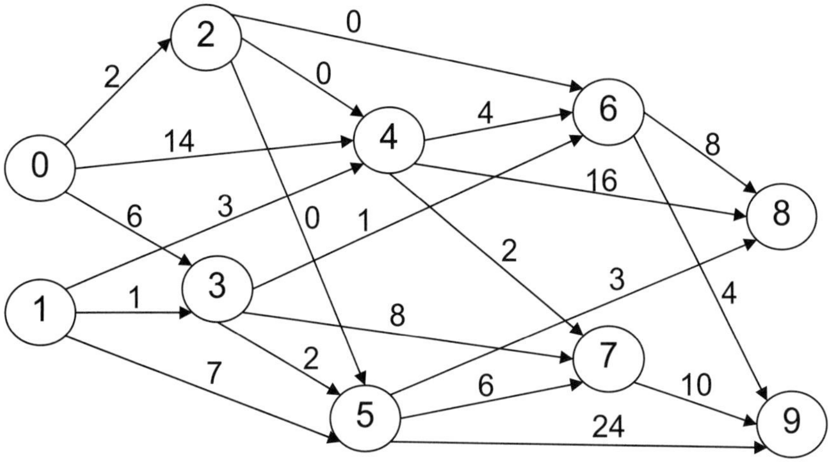

Fig. 4.10. Scheduling the kth task in case (i).

$$g_1^*(t, \lambda) = \min_{i \in J_n} \{g_1^*(t - p_i, \lambda) + f_i(t) + \lambda_i\} \tag{4.55}$$

Also in this case we obtain a lower bound to the schedule cost.

The third improvement is obtained by introducing a non-negative integer q_i, $i = 1, \ldots, n$, called the state space *modifier*. The following recurrence relation is formulated in this case:

$$g_2^*(t, q) = \min_{i \in J_n} \{g_2^*(t - p_i, q - q_i) + f_i(t)\} \tag{4.56}$$

initialized by $g_2^*(t, q) = \infty$ for $t < 0$ or $q < 0$, and $g_2(0, 0) = 0$. The computation of $g_2^*(\sum_{i=1}^n p_i, \sum_{i=1}^n q_i)$ requires $O(n \sum_{i=1}^n p_i(1 + \sum_{i=1}^n q_i))$ time.

The branch and bound algorithm examined in [1] uses the lower bounds defined above. A node at level l of the branching tree corresponds to a partial schedule of the first k tasks. An adjacent task interchange is applied at each node (except for the first level) in order to eliminate the dominated nodes. The lower bounds are calculated for the nodes that cannot be eliminated. Abdul-Razaq and Potts in [1] present results of computational experiments performed to compare the proposed lower bounds. Although the proposed lower bounds are quite tight, their computation is too time consuming. The algorithm solves instances with up to 25 nodes in 100 seconds. No one of the compared lower bounds proved to be significantly superior over the others.

Ow and Morton in [203] propose a filtered beam search algorithm to solve the problem with objective function (4.1) and the no idle time constraint. They prove a condition that allows to establish the order of two adjacent tasks in the sequence. The condition is formulated below as Property 4.14.

Property 4.14 *All adjacent pairs of tasks in the optimal sequence must satisfy the following condition*

$$\beta_i p_i - \Omega_{ij}(\beta_i + \alpha_i) \geq \beta_i p_i - \Omega_{ij}(\beta_i + \alpha_i) \tag{4.57}$$

where task i immediately precedes task j, and Ω_{ij} and Ω_{ji} are defined as

$$\Omega_{ij} = \begin{cases} 0 & \text{if} \quad d_k - t - p_k \leq 0 \\ d_k - t - p_k & \text{if} \quad 0 < d_k - t - p_k < p_k \\ p_k & \text{otherwise} \end{cases} \tag{4.58}$$

where t is the earliest time the machine is free.

Ow and Morton show that Properties 4.1 and 4.2 also hold under the no idle time constraint. Moreover, they propose priority rules given in formulas (4.59) and (4.60):

$$o_i^1 = \begin{cases} \dfrac{\beta_i}{p_i} & \text{if } y_i \leq 0 \\ \dfrac{\beta_i}{p_i} - y_i \dfrac{(\beta_i + \alpha_i)}{p_i k \overline{p}} & \text{if } 0 \leq y_i \leq k\overline{p} \\ -\dfrac{\alpha_i}{p_i} & \text{otherwise} \end{cases} \tag{4.59}$$

$$o_i^2 = \begin{cases} \dfrac{\beta_i}{p_i} & \text{if } y_i \leq 0 \\ \dfrac{\beta_i}{p_i} exp\left(-\dfrac{\alpha_i + \beta_i}{\alpha_i}\dfrac{y_i}{\overline{p}}\right) & \text{if } 0 \leq y_i \leq \left(\dfrac{\beta_i}{\alpha_i + \beta_i}\right) k\overline{p} \\ \dfrac{1}{\alpha_i^2}\left(\dfrac{\beta_i}{p_i} - \dfrac{(\alpha_i + \beta_i)(y_i)}{k\overline{p}}\right)^3 & \text{if } \dfrac{\beta_i}{\alpha_i + \beta_i} k\overline{p} \leq y_i \leq k\overline{p} \\ -\dfrac{\alpha_i}{p_i} & \text{otherwise} \end{cases} \tag{4.60}$$

where $y_i = d_i - t - p_i$, $\overline{p} = \frac{1}{n}\sum_{i=1}^{n} p_i$, and k is a control parameter indicating the point at which tardiness is imminent. Before this time only the earliness cost is relevant. Large values of k are suggested in the case of close due dates, and small values in the case of evenly distributed due dates and short processing times.

The basic concept of the beam search algorithm introduced in [202] is to search a limited number of solution paths in parallel. Each node in the beam search tree corresponds to a partial sequence of tasks. Each descendant of a node is obtained by appending one unscheduled task to the sequence. Each partial schedule is evaluated using the *evaluation function.* The filtered beam search algorithm developed by Ow and Morton uses two parameters, *filterwidth* and *beamwidth.* First, all nodes are evaluated by the *priority function.* It is equal to the priority of the last added task. The nodes with the highest priority pass to the next step, which is the calculation of the *evaluation function*, i.e. the cost of a partial schedule. In order to calculate this cost the partial schedule is completed with the remaining tasks, in the order following from some priority rule. The cost of the obtained schedule is an upper bound to the optimal schedule cost. The filterwidth determines the number of nodes for which the evaluation function is calculated. The best nodes are selected for further exploration. The beamwidth determines the number of nodes that are further explored. No backtracking is used. Obviously, beam search is a heuristic algorithm. The computational results presented in [203] show that the algorithm is quite accurate and significantly superior to simple priority driven algorithms based on the EDD rule or on priorities given by formulas (4.59) and (4.60).

Li in [173] proposes a branch and bound algorithm and a heuristic to solve the problem with task dependent due dates and no idle time inserted, with minimization of the total earliness and tardiness cost. The problem is decomposed into two simpler subproblems one with minimization of the total tardiness cost, and in the other with minimization of the total earliness cost. Lower bounds are calculated separately for each subproblem and the following property is used to calculate the lower bound to the earliness/tardiness cost.

Property 4.15 *If LB_t is the lower bound to the total tardiness cost, and LB_e is the lower bound on the total earliness cost, then $LB_e + LB_t$ is a lower bound on the total earliness and tardiness cost.*

The lower bounds for the subproblems are calculated using the Lagrangian relaxation. A node at level l of the branching tree corresponds to a partial sequence with the order of l tasks fixed. If the unit tardiness cost is large, the sequence is constructed starting from the first position forward. If the tardy factor is small, the sequence is constructed starting from the last position backward. Three test are performed to decide whether a node should be fathomed or not. The tests consist in

- verifying if the sequence obtained by adding task i does not violate the order imposed by Properties 4.1 and 4.2;
- interchanging the last two tasks; if the interchange results in a better partial schedule then the node is discarded;
- applying the dominance principle of dynamic programming.

If the above tests do not lead to discarding the node, then the lower bound for the node is calculated. The tree is searched using the depth-first strategy. The initial upper bound is obtained by a local search heuristic given as Algorithm 4.16 below. The interesting feature of the heuristic is that the operator used to generate the neighborhood changes during the search process. The operator interchanges two tasks separated by j tasks, where j is the operator parameter.

Two types of the stopping condition are considered. The algorithm stops either if it the limit of computational time is achieved or if, despite changing the operator, the process remains trapped in a local optimum. The parameter z is chosen empirically and depends on the instance of the scheduling problem. Generally, if due dates are tight, z should be large.

Algorithm 4.16 (Li [173]).

1. Create an initial solution.
2. Start with the first or with the last task in the sequence and apply the *current operator* to generate the neighborhood of the current solution.
3. If there are solutions in the neighborhood that have a lower value of the total earliness and tardiness cost than the current solution, then select one such solution, make it the current solution and go to step 2.
4. If the stopping condition is met then stop, else change to another operator, i.e. if $j < z$ then set $j = j + 1$, else set $j = 0$ and go to step 2.

As reported in [173], the branch and bound algorithm solves instances with up to 50 tasks in 100 seconds. The heuristic solves instances with even 3000 tasks.

Another approach to solving the earliness/tardiness problem with task dependent due dates and no idle time is proposed by Liaw in

[175]. The lower bounds are obtained by the Lagrangian relaxation. The following property is proposed in [175] to determine the optimal order of non-adjacent tasks.

Property 4.16 *All pairs of non-adjacent tasks in an optimal schedule must satisfy the following condition:*
if $p_i = p_j$, *then* $\beta_i(p_j + \Delta) - \Lambda_{ij}(\beta_i + \alpha_i) \geq \beta_j(p_i + \Delta) - \Lambda_{ji}(\beta_j + \alpha_j)$ *where task* i *precedes task* j *and* Λ_{ij} *is defined as follows*

$$\Lambda_{ij} = \begin{cases} 0 & \text{if } d_i - t - p_i \leq 0 \\ d_i - t - p_i & \text{if } 0 < d_i - t - p_i < p_j + \Delta \\ p_j + \Delta & \text{otherwise} \end{cases} \tag{4.61}$$

where Δ *is the sum of processing times of all tasks between task* i *and task* j, *and* t *is the sum of processing times of all tasks preceding task* i.

The initial upper bound for the branch and bound algorithm is obtained by the following heuristic algorithm.

Algorithm 4.17 (Liaw [175]).

1. Find an initial sequence using priorities defined by formula (4.60).
2. (Insertion procedure) For all tasks $i = 1, \ldots, n$, and for all tasks k in positions $i - \lfloor n/3 \rfloor, \ldots, i - 1$, and $i + 1, \ldots, i + \lfloor n/3 \rfloor$ in the sequence, obtain a new sequence by inserting task k immediately before task i. Choose the best one from all sequences obtained in this way.
3. (Interchange procedure) For all tasks $i = 1, \ldots, n$ and for all tasks k in positions $i - \lfloor n/3 \rfloor, \ldots, i - 1$ and $i + 1, \ldots, i + \lfloor n/3 \rfloor$ in the sequence, obtain a new sequence by interchanging tasks i and k. Choose the best one from all sequences obtained in this way.

In the implementation of the branch and bound algorithm, a node at level l corresponds to a partial schedule with tasks in the last l positions fixed. The depth-first strategy is used to search the tree. The first two tests used also in [173] are applied to eliminate a node. If the node cannnot be eliminated, a lower bound is calculated for this node. If the lower bound plus the cost of the partial schedule is larger than the current upper bound, then the node is discarded.

The computational experiments reported in [175] show that problems with 50 tasks can be solved by the branch and bound algorithm within one hour of computation. The lower bounds proved slightly tighter that those proposed in [173], but average computational times for the corresponding instances are longer.

Nandkeolyar et al. in [194] propose several heuristic algorithms to solve the problem with arbitrary task ready times. They compare the behavior of the heuristics for various levels of machine utilization on a basis of a computational experiment.

4.2.2 Task independent weights

Szwarc in [233] considers a special case of the earliness/tardiness problem with task dependent due dates and no idle time inserted, where the unit earliness and tardiness costs are task independent, i.e. $\alpha_i = \alpha$ and $\beta_i = \beta$, $i = 1, \ldots, n$. The problem is NP-hard since the minimization of the total tardiness is NP-hard (see [92]). Szwarc proves several properties of optimal schedules and develops a branching scheme that uses the decomposition rules following from these properties. The computational experiment shows that instances with up to 10 tasks may be solved optimally without applying any lower bounds.

In order to introduce the properties proved in [233], let us consider a pair of adjacent tasks i, j. The cost of a schedule obtained by interchanging tasks i and j is a linear function of the start time t of the first of the two tasks. It may be proved that for each pair of tasks i and j, there exists a critical value t_{ij} such that task i precedes task j in an optimal schedule if $t \geq t_{ij}$, and j precedes i if $t < t_{ij}$. In fact, there may exist an interval in which both orderings are optimal, but in that case we assume that t_{ij} coincides with the beginning of the interval. Let us arrange tasks according to non-decreasing processing times, breaking ties in favor of the task with the shorter due date. The critical values t_{ij} for $i < j$ are calculated as follows.

$$t_{ij} = \begin{cases} d_i - \dfrac{\alpha p_i + \beta p_j}{\alpha + \beta} & \text{if } d_i - d_j > \dfrac{\alpha}{\alpha + \beta}(p_i - p_j) \\ d_i - p_i - \dfrac{\alpha p_i + \beta p_j}{\alpha + \beta} & \text{if } d_i - d_j \leq \dfrac{\alpha}{\alpha + \beta}(p_i - p_j) \\ 0 & \text{if } p_i = p_j \end{cases} \tag{4.62}$$

Based on the critical values defined above, a parametric precedence matrix is constructed. If task i precedes task j unconditionally (i.e. for all t) the relevant entry is marked "$-$"; if task j precedes i unconditionally, the entry is marked "$+$". Otherwise, the entry takes the value t_{ij}. Substantial reduction of the computational effort necessary to find an optimal sequence can be achieved by decomposing the problem into smaller subproblems. Szwarc proposes the following decomposition rules, presented as Properties 4.17 - 4.19.

Property 4.17 ([233]) *(a) Task i is ordered first in an optimal schedule if all entries of the precendence matrix are "+" in column i and "−" in row i.*
(b) Task i is ordered last in an optimal schedule if all entries of the precendence matrix are "−" in column i and "+" in row i.

The scheduling problem can be decomposed into separate subproblems. Each subproblem involves a subset of tasks, called a *block*, with fixed start and completion time. Each block is scheduled independently, and the final schedule is obtained as a concatenation of the schedules of consecutive blocks.

Property 4.18 *Block $A = (1, \ldots, m)$ precedes block $B = (m+1, \ldots, n)$ in an optimal schedule if all entries (i, j) are "−" for each $i \leq m$ and each $j \geq m + 1$.*

Let us assume that all entries in row i and column i are "+" or "−". Let A and B such that $A \cup B \cup \{i\} = J$ be nonempty sets where j precedes i unconditionally for all $j \in A$ and i precedes k unconditionally for all $k \in B$. Then the following property holds.

Property 4.19 *If $p_i \leq \max\{\min_{j \in A}\{p_j\}, \min_{k \in B}\{p_k\}\}$ then AiB is the optimal block schedule.*

The branching strategy proposed in [233] generates schedules that cannot be improved by adjacent task interchanges. Each node at level l in the branching tree corresponds to a partial sequence with the order of the first l tasks fixed. A descendant of a node is obtained by adding an unscheduled task at the end of the sequence represented by the parent node. The branching scheme considers only nodes created by adding task i such that for some unscheduled task j, task i precedes task j for t calculated as the total processing time of the l scheduled tasks.

4.3 Controllable due dates

Task dependent due dates are in some circumstances negotiable. On one hand a long due date allows completing the order on time. On the other hand the customer may wish to receive a better price if he has to wait longer. Thus we may assume that delaying the due date incurs an additional cost. The earliness/ tardiness scheduling problem with controllable task dependent due dates is formulated as follows. Consider n independent, nonpreemptive tasks with processing times p_i, unit earliness costs α_i, unit tardiness costs β_i, and unit due date cost γ_i, $i = 1, \ldots, n$. The problem is to find task due dates and a schedule of tasks on a single machine, such that the following objective function is minimized.

$$\sum_{i=1}^{n} (\alpha_i e_i + \beta_i t_i + \gamma_i d_i) \tag{4.63}$$

Surveys of scheduling problems with task dependent controllable due dates can be found in [60] and [107].

Seidmann et al. in [216] formulate a special case of this problem. They consider the situation where the customer accepts a given lead time A of an order without additional cost. Therefore only setting a due date longer than A incurs additional costs, proportional to the value $\max\{0, d_i - A\}$. The objective function is given by formula (4.64).

$$\sum_{i=1}^{n} (\alpha e_i + \beta t_i + \gamma \max\{0, d_i - A\}) \tag{4.64}$$

The following polynomial time algorithm solving the above problem is proposed in [216].

Algorithm 4.18 (Seidmann et al. [216]).

1. If $\gamma \le \beta$ then set $d_i^* = \sum_{j=1}^{i} p_j$, $i = 1, \ldots, n$, else set $d_i^* = \min\{A, \sum_{j=1}^{i} p_j\}$.
2. Schedule tasks according to the shortest processing time.

It is easy to notice that optimal schedules do not contain any idle time.

Let us illustrate the algorithm with the following example.

Example 4.19. Consider $n = 6$ tasks with processing times $p_1 = 4$, $p_2 = 7$, $p_3 = 9$, $p_4 = 10$, $p_5 = 11$, $p_6 = 14$, penalties $\alpha = 24$, $\beta = 21$, $\gamma = 32$, and $A = 30$. Since $\gamma > \beta$, the optimal due dates obtained in step 1 of Algorithm 4.18 are $p_1 = 4$, $p_1 = 11$, $p_3 = 20$, $p_4 = 30$, $p_5 = 30$, $p_6 = 30$. Total cost, which is equal to the tardiness costs of tasks 5 and 6, equals 525.

4.3.1 TWK due date model

The problems where due dates are defined as $d_i = r_i + vp_i$, $i = 1, \ldots, n$ is called the TWK (total work content) problem, where r_i is the ready time of task i, and v is a variable. Cheng in [50] (see also [193]) examines the problem of minimization of the total squared deviation of task completion times from task dependent due dates defined as $d_i = vp_i$, where v is a decision variable, $i = 1, \ldots, n$. It is easy to notice that this is a special case of the TWK problem with $r_i = 0$, $i = 1, \ldots, n$. The objective function is given by formula (4.65).

$$\sum_{i=1}^{n} (C_i - vp_i)^2 \tag{4.65}$$

The problem is to find a sequence of tasks and the variable v minimizing function (4.65). Cheng gives a polynomial time algorithm to solve the problem. The algorithm is presented below.

Algorithm 4.20 (Cheng [50]).

1. Schedule tasks according to the shortest processing time.
2. Set
$$v^* = \frac{\sum_{i=1}^{n} p_i \sum_{j=1}^{i} p_j}{\sum_{i=1}^{n} p_i^2}$$

Let us illustrate the algorithm with an example.

Example 4.21. Consider $n = 4$ tasks with processing times $p_1 = 4$, $p_2 = 7$, $p_3 = 9$, and $p_4 = 10$. The optimal value of v is calculated as

$$v^* = \frac{4 * 4 + 7(4 + 7) + 9(4 + 7 + 9) + 10(4 + 7 + 9 + 10)}{16 + 49 + 81 + 100} = 2.33$$

We obtain the following optimal due dates: $d_1^* = 9.32$, $d_2^* = 16.31$, $d_3^* = 20.96$ and $d_4^* = 23.29$, and the total cost equals 102.33.

Gupta et al. in [113] consider the symmetric TWET problem with TWK due dates. They prove the following property useful for finding optimal schedules.

Property 4.20 *For any given sequence of tasks there exists an optimal value v^* which coincides with the ratio of completion time and processing time of exactly one task in the sequence.*

Cheng in [53] examines the TWK problem with due dates defined as $d_i = vp_i^m$, $i = 1, \ldots, n$, where m is given and v is a decision variable. The objective is the minimization of function (4.66).

$$\gamma v + \sum_{i=1}^{n} (e_i + t_i) \tag{4.66}$$

where γ is a unit cost of v. The problem is formulated as a linear programming problem and solved in $O(n^2)$ time ([53]). A simpler algorithm of complexity $O(nlogn)$ for this problem is proposed by van de Velde in [240].

Chu and Gordon prove in [68] that if unit earliness and tardiness costs are task dependent, i.e. the objective is minimization of function (4.67), the problem becomes NP-hard, even if weights are symmetric.

$$\gamma v + \sum_{i=1}^{n} (\alpha_i e_i + \beta_i t_i) \tag{4.67}$$

As shown in [68], in the special case, when $\gamma = 0$ the problem can be solved in $O(nlogn)$ time using the algorithm proposed in [123] for the common due date problem with equal processing times of tasks. This algorithm is presented in Chapter 3 as Algorithm 3.34.

The computational complexity of the earliness/tardiness problem with TWK due date model is summarized in Table 4.4.

Table 4.4. Computational complexity of earliness/tardiness problems with TWK due date model

Objective function	Complexity	Algorithm
$\gamma v + \sum_{i=1}^{n} (e_i + t_i)$	P	$O(nlogn)$, [53, 240]
$\gamma v + \sum_{i=1}^{n} \alpha_i (e_i + t_i)$	NP-hard, [68]	
$\sum_{i=1}^{n} (\alpha_i e_i + \beta_i t_i)$	P	$O(nlogn)$, [68]
$\sum_{i=1}^{n} (e_i^2 + t_i^2)$	NP-hard [107]	heurist. [171, 139]
		enumer. [112, 232]

4.3.2 SLK due date model

Another problem with task dependent due dates, where due dates are given as $d_i = r_i + p_i + q$, $i = 1, \ldots, n$, is called the SLK (slack) problem. We assume that $r_i = 0$, $i = 1, \ldots, n$. In general, the objective is to minimize the total weighted earliness and tardiness cost. Let us first consider a special case of the SLK problem with symmetric earliness and tardiness penalties. Taking into account that

$$|C_i - d_i| = |C_i - p_i - q| = |C_{i-1} - q|, \; i = 1, \ldots, n, \tag{4.68}$$

where $C_0 \geq 0$, the problem may be formulated as the following linear programming problem.

$$\text{minimize} \; \sum_{i=1}^{n} \alpha_i |C_{i-1} - q| \tag{4.69}$$

subject to

$$C_0 \geq 0 \tag{4.70}$$

$$C_i - C_{i-1} \geq p_i, i = 1, \ldots, n. \tag{4.71}$$

This formulation is very similar to the formulation of the common due date problem with symmetric weights (3.16) considered in Chapter 3. Gupta et al. in [113] observe this similarity and prove the following properties of the SLK problem with symmetric weights. Similarly as for the common due date problem, there exists a V-shaped optimal schedule.

Property 4.21 *There exists an optimal schedule for the problem (4.69) which is V-shaped in the sense that the non-tardy tasks (set E) are sequenced in the non-increasing order of the ratio p_i/α_i and followed by the tardy tasks (set T) sequenced in the non-decreasing order of the ratio p_i/α_i.*

Moreover, since q is a decision variable the completion time of one task coincides with the optimal value of q.

Property 4.22 *There exists an optimal schedule for the problem (4.69) such that the optimal slack q^* coincides with the start time of the last non-tardy task.*

Finally, the sum of weights of tardy tasks does not exceed the sum of weights of non-tardy tasks.

Property 4.23 *There exists an optimal schedule for the problem (4.69) such that* $\sum_{i\in T}\alpha_i \le \sum_{i\in E}\alpha_i$.

Using Properties 4.21 - 4.23 a polynomial time algorithm for the case with $\alpha_i = 1$, $i = 1, \ldots, n$, similar to Algorithm 3.1, may be constructed. For the same problem Karacapilidis and Pappis in [146, 147] propose an algorithm to find an optimal value of q and all optimal sequences. Each sequence is found in $O(nlogn)$ time.

The SLK problem is considered restricted if

$$q < \begin{cases} p_1 + p_3 + \ldots + p_{n-1} & \text{if } n \text{ is even} \\ \\ p_2 + p_4 + \ldots + p_{n-1} & \text{if } n \text{ is odd} \end{cases} \tag{4.72}$$

assuming that tasks are ordered according to non-decreasing processing times. Oguz and Dincer in [200] prove that the restricted problem is NP-hard even if $\alpha_i = 1$, $i = 1, \ldots, n$.

Gupta et al. in [113] present a heuristic algorithm for the SLK problem with arbitrary weights α_i, $i = 1, \ldots, n$.

Adamopoulos and Pappis in [3] consider the SLK problem with the objective function given by formula (4.73) with $\alpha_i = \lambda p_i^a$, and $\beta_i = p_i^b$, $i = 1, \ldots, n$, where $\lambda > 0$, and parameters a and b are non-negative integers.

$$\sum_{i=1}^{n} (\alpha_i e_i + \beta_i p_i) \tag{4.73}$$

The following cases are distinguished:

- $a = b > 1$;
- $a = 1$, $b > 1$ or $a > 1$, $b = 1$;
- $a > 1$, $\beta_i = 1$ or $b > 1$, α_i, $i = 1, \ldots, n$;
- $a = 1$, $\beta_i = 1$ or $b = 1$, α_i, $i = 1, \ldots, n$.

Properties of optimal schedules and branch and bound algorithms for each case are given in [3]. For a given sequence the task that starts at time q^* can be found using the following property.

Property 4.24 *For any specified sequence in which the optimal slack* q^* *coincides with the starting time of task* k, k *is determined by:*

$$\sum_{i=1}^{r-1} \alpha_i - \sum_{i=r}^{n} \beta_i < 0 \tag{4.74}$$

$$\sum_{i=1}^{r} \alpha_i - \sum_{i=r+1}^{n} \beta_i < 0 \tag{4.75}$$

Since the optimal position k is not known before the sequence is fixed, the algorithm examines all values $1, \ldots, n$. The algorithm developed in [3] for the case $a = 1$, $\beta_i = 1$ or $b = 1$, α_i, $i = 1, \ldots, n$ is presented below.

Algorithm 4.22 (Adamopoulos and Pappis [3]).

1. Order tasks according to non-decreasing processing times. Set $l = 1$.
2. If $l > n$ then go to step 5. Assign tasks scheduled before position l to set E, and tasks scheduled starting from position l to set T. Construct sequence S by ordering tasks in sets E and T according to the V-shape property. Determine position k according to Property 4.24. If $l \neq k$ then go to step 4. Calculate the cost of schedule S.
3. Interchange positions of two tasks (one in set E and one in set T) so that the V-shape is not violated. Determine position k' for the new sequence S'. If $k' \neq l$ then go to step 4. Calculate the cost of schedule S'. If the cost of sequence S' is smaller then store sequence S' and repeat step 3, else store sequence S and repeat step 3.
4. Set $l = l + 1$ and go to step 2.
5. The optimal sequence is the one that corresponds to the minimum cost.

The remaining algorithms proposed in [3] differ in step 2 where the order of tasks in sets E and T is determined. Let us illustrate the algorithm with an example.

Example 4.23. Given is a set of $n = 7$ tasks with processing times $p_1 = 4$, $p_2 = 5$, $p_3 = 7$, $p_4 = 9$, $p_5 = 11$, $p_6 = 14$, and $p_7 = 17$, and $a = b = 2$. The corresponding weights are then $\alpha_i = \beta_i = p_i^2$. The sequences examined in consecutive iterations are presented in Table 4.5.

Kahlbacher in [138] proves that the following problems are equivalent under some conditions imposed on functions g, h and the due dates:

(a) the problem to minimize the objective function $\sum_{i=1}^{n} g(C_i - d)$ with a common due date;

Table 4.5. Sequences examined by Algorithm 4.22 in Example 4.23

$l=1$	7,6,5,4,3,2,1 $k=2$				
$l=2$	1,7,6,5,4,3,2 $k=3$				
$l=3$	1,2,7,6,5,4,3 $k=4$				
$l=4$	1,2,3,7,6,5,4 $k=5$				
$l=5$	1,2,3,4,7,6,5 $k=5$ $f(S)=9521$	$\rightarrow$ 1,2,3,5,7,6,4 $k=5$ $f(S)=9063$	$\rightarrow$ 1,2,3,6,7,5,4 $k=5$ $f(S)=9228$		
			$\rightarrow$ 1,2,4,5,7,6,3 $k=5$ $f(S)=$ **8891**	$\rightarrow$ 1,2,4,6,7,5,3 $k=5$ $f(S)=9248$	
$l=6$	1,2,3,4,5,7,6 $k=6$ $f(S)=8982$	$\rightarrow$ 1,2,3,4,6,7,5 $k=6$ $f(S)=9633$			
$l=7$	1,2,3,4,5,6,7 $k=6$				

(b) the problem to minimize the objective function $\sum_{i=1}^{n} h(C_i - d_i)$ with $d_i = p_i + q$, $i = 1, \ldots, n$.

The conditions to be met are the following:

- $q = \sum_{i=1}^{n} p_i - d$;
- $h(x) = g(-x)$ and $g(x)$ is an unimodal, real valued function with the following properties:
  - $g(0) = 0$;
  - $g(x_1) \geq g(x_2)$ for all $x_1 \leq x_2 \leq 0$;
  - $g(x_1) \leq g(x_2)$ for all $0 \leq x_1 \leq x_2$.

It is worth noticing that objective functions of the MAD, WSAD and MSD problems satisfy these conditions. Therefore, in view of this observation the NP-hardness of the MSD problem with SLK due dates follows from the fact that the MSD problem with a common due date is NP-hard (see [153]).

Finally, let us consider the problem of minimizing the objective function (4.76).

$$\sum_{i=1}^{n} (\alpha e_i + \beta t_i + \gamma q) \tag{4.76}$$

A polynomial time algorithm solving this problem is proposed by Adamopoulos and Pappis in [4]. First the position k of the last early task is calculated as $k = \lceil n(\beta - \gamma)/(\alpha + \beta) \rceil$. Then, position weights are calculated and tasks are allocated in such positions that the pairwise product of the positional weights and task processing times is minimized. The optimal slack is calculated as $q^* = C_k - p_k$ if tasks are indexed according to the position in the optimal schedule.

Cheng et al. in [66] consider the problem with controllable processing times of tasks. It is assumed that the normal processing time p_i of task i, $i = 1, \ldots, n$, can be reduced at cost λ_i per time unit, but not more than to the value p_i^{min}. Thus, if task i is compressed by x_i time units its processing cost equals $x_i \lambda_i$, $i = 1, \ldots, n$, $0 \leq x_i \leq (p_i - p_i^{min})$. The objective is to find a set of processing times and the sequence of tasks to minimize the cost function (4.77).

$$\sum_{i=1}^{n} (\alpha e_i + \beta t_i + \gamma q + \lambda_i x_i) \tag{4.77}$$

The following properties are proved for the SLK due dates and controllable processing times in [66].

Property 4.25 *If $\beta < \gamma$ then $q^* = 0$.*

Property 4.26 *For given $x_1, \ldots, x_n$, in an optimal sequence the last early task is the shortest task.*

Property 4.27 *For given $x_1, \ldots, x_n$, in any sequence there exists an optimal value of q such that the last early task completes at its due date.*

Property 4.28 *For given $x_1, \ldots, x_n$, in any sequence there exists an optimal value of q equal to the completion time of the kth task in the sequence, where $k = \lceil n(\beta - \gamma)/(\alpha + \beta) \rceil$.*

Assume that the sequence of tasks is given and the tasks are indexed according to their positions in the sequence. Then the objective function (4.77) can be rewritten as follows:

$$\sum_{i=1}^{k-1}(\alpha_i+\gamma(n+1)-\lambda_i)p_i+\sum_{i=k}^{n}(\beta(n-i)+\gamma-\lambda)p_i+\sum_{i=1}^{n}\lambda_i p_i \quad (4.78)$$

where k is the last non-tardy task. Thus the position weights are defined as follows:

$$w_i=\begin{cases}\alpha_i+\gamma(n+1)-\lambda_i, & 1\leq i\leq k-1,\\ \beta(n-i)+\gamma-\lambda_i, & k\leq i\leq n.\end{cases} \quad (4.79)$$

The optimal processing times of tasks can be found on the basis of the following property proved in [66].

Property 4.29 *Given a sequence for the SLK problem, the optimal processing times can be determined as follows:*

- *the optimal processing time of a task with $w_i<0$ equals p_i;*
- *the optimal processing time of a task with $w_i>0$ equals p_i^{min};*
- *the optimal processing time of a task with $w_i=0$ can be any value in the interval $[p_i^{min},p_i]$.*

Now, the scheduling problem may be formulated as the following assignment problem.

$$\text{minimize} \quad \sum_{i=1}^{n}\sum_{j=1}^{n}c_{ij}y_{ij} \quad (4.80)$$

subject to

$$\sum_{i=1}^{n}y_{ij}=1, j=1,\ldots,n, \quad (4.81)$$

$$\sum_{j=1}^{n}y_{ij}=1, i=1,\ldots,n, \quad (4.82)$$

$$y_{ij}\in\{0,1\}, i=1,\ldots,n, j=1,\ldots,n, \quad (4.83)$$

where

$$c_{ij}=w_{ij}(p_{ij}-x_{ij}), i=1,\ldots,n, j=1,\ldots,n, \quad (4.84)$$

$$w_{ij}=\begin{cases}\alpha_i+\gamma(n+1)-\lambda_j, & 1\leq i\leq k-1,\\ \beta(n-i)+\gamma-\lambda_j, & k\leq i\leq n.\end{cases} \quad (4.85)$$

and

$$x_{ij} = \begin{cases} p_i & \text{if } w_{ij} < 0, \\ p'_i & \text{if } w_{ij} = 0, \\ p_i^{min} & \text{if } w_{ij} > 0, \end{cases} \tag{4.86}$$

where

$$p_i^{min} \le p'_i \le p_i. \tag{4.87}$$

The optimal solution of the assignment problem defines an optimal solution of the scheduling problem, where $y_{ij} = 1$ means that task i is assigned to position j in the optimal sequence. Recall that the assignment problem can be solved in time $O(n^3)$.

For a special case of the problem (4.77) where $\lambda_i = \lambda$ and $x_i = x$, the following algorithm of complexity $O(nlogn)$ is presented in [66].

Algorithm 4.24 (Cheng et al. [66]).

1. Set $k = \max\{0, \lceil n(\beta - \gamma)/(\alpha + \beta)\rceil\}$.
2. Calculate w_j, $j = 1, \ldots, n$, according to formula (4.79).
3. Arrange the weights in the non-increasing order.
4. Find the optimal sequence of tasks by matching the tasks in non-decreasing order of normal processing times p_i with the weights in non-increasing order.
5. Determine the optimal processing times of tasks according to Property 4.29.
6. Determine the optimal slack q^* as the sum of optimal processing times of tasks scheduled in positions 1 through $k - 1$.

The computational complexity of the earliness/tardiness problem with SLK due date model is summarized in Table 4.6.

4.3.3 Scheduling with batch setup times

Another problem with controllable due dates is considered by Chen in [48]. In this problem tasks are processed in batches. All tasks in a batch have the same due date and there is a machine setup required between batches. The problem is formulated as follows. Lets us consider $b \ge 1$ batches of nonpreemptive and independent task to be processed on a single machine. Each batch consists of n_j tasks and has a due date d_j, $j = 1, \ldots, b$. Morover, γ_j is the unit cost incurred by the due date d_j. A setup time s_{lj} is required whenever a task from batch j is processed immediately after a task from batch l, $j = 1, \ldots, b$, $l = 1, \ldots, b$. Let

Table 4.6. Computational complexity of earliness/tardiness problems with SLK due date model

Objective function	Complexity	Algorithm
$\sum_{i=1}^{n}(e_i + t_i)$	P	$O(nlogn)$, [113, 146, 147]
$\sum_{i=1}^{n}(e_i + t_i)$ (restricted)	NP-hard, [200]	
$\sum_{i=1}^{n}(\alpha e_i + \beta t_i + \gamma q)$	P	$O(nlogn)$, [4]
$\sum_{i=1}^{n}\alpha_i(e_i + t_i)$	NP-hard	heurist. [113]
$\sum_{i=1}^{n}(e_i^2 + t_i^2)$	NP-hard [107]	

us denote by p_{ij} the processing time, and by C_{ij} the completion time of task i in batch j. We define the earliness and tardiness of task i in batch j as follows:

$$e_{ij} = \max\{0, d_i - C_{ij}\}, \tag{4.88}$$

$$t_{ij} = \max\{0, C_{ij} - d_i\}. \tag{4.89}$$

The unit earliness cost α_j and tardiness cost β_j are equal for all tasks in batch j, $j = 1, \ldots, n$. The problem is to find a schedule with minimum total earliness/tardiness and due date cost given by (4.50).

$$\sum_{j=1}^{b}\sum_{i=1}^{n_j}(\alpha_j e_{ij} + \beta_j t_{ij}) + \sum_{j=1}^{b}\gamma_j d_j \tag{4.90}$$

Properties of optimal schedules of problem 4.90, summarized below as Property 4.30 are proved in [48].

Property 4.30 *An optimal schedule has the following properties.*

1. *An optimal schedule starts at time zero and contains no idle time.*
2. *Tasks from batch j form a V-shape with respect to their due date d_j, $j = 1, \ldots, n$.*
3. *If $n_j\beta_j \le \gamma$, then the optimal due date for batch j equals zero, otherwise task i_j, $i_j = \lceil (n_j\beta_j - \gamma_j)/(\alpha_j + \beta_j) \rceil$, completes exactly at the due date, i.e. $C_{i_j j} = d_j$.*

In the case of two batches the problem is solvable in $O(n^8)$ time by a dynamic programming algorithm proposed in [48]. The complexity of the problem with $b \geq 3$ remains an open question. However, if all batches share the same due date, i.e. $d_j = d$, $j = 1, \ldots, b$, the problem can be solved in $O(n^{2b}/b^{2b-3})$ time by a dynamic programming algorithm presented in [48].

5

Algorithms for schedule balancing

The idea of balancing the schedules in order to meet the just-in-time objectives originated in Toyota. Generally speaking, it is required that for every part type the number of units of this part used for assembly remains constant in each time period (*stage*). The order of products scheduled on the assembly line allows to determine the number of parts required for assembly, based on the explosion of the bill of material. Thus, the variation of the production volume at the part manufacturing level can be minimized by appropriate scheduling of the finished products on the assembly line.

Monden in [191] gives the first description of the Toyota production control system. The first mathematical model of the Totota system is proposed by Miltenburg in [186]. Miltenburg defines two basic production systems: the multi-level and the single-level system. The scheduling problems formulated for both systems consider various optimality criteria. In general, the objective is to minimize the maximum or total cost of deviation of the scheduled production volume from the ideal volume.

In this chapter we introduce the single-level as well as the multi-level system and examine the main optimality criteria considered in the literature. We also present algorithms developed to find balanced schedules for various optimality criteria. Finally, we describe a transformation of the single-level just-in-time scheduling problem to the problem of apportionment, introduced in Section 2.2 and we present the classification of the scheduling algorithms based on this transformation.

The last section of this chapter is devoted to the hard real-time scheduling problems. We formulate the Liu-Layland problem and scheduling algorithms presented in [177]. Finally, we show the transformation of the Liu-Layland problem to the problem of apportionment and discuss the possibility of using the methods of apportionment to solve the Liu-Layland problem.

In Section 5.1 a multi-level just-in-time scheduling problem is defined and analyzed, while in Section 5.2 the single-level problem is examined. Finally, various approaches to solving the Liu-Layland problem are discussed in Section 5.3.

5.1 The multi-level scheduling problem

Most of the manufacturing processes (or supply chains) consist of several production levels. It means that raw materials or purchased parts are first fabricated into components which are then combined into sub-assemblies. Finally, sub-assemblies are assembled into products on an assembly line. The structure of a product is often represented as a so-called gozintograph or a bill of material. In Section 1.1.3 we introduced the concept of the bill of material and the multi-level structure of a product. An example of a four-level product structure is presented in Fig.1.2.

The scheduling problem considered in this section was first fomulated by Miltenburg and Sinnamon in [188] as the muli-level mixed-model just-in-time scheduling problem. Later, Kubiak ([153]) proposed the name *Output Rate Variation Problem* (ORV).

5.1.1 Problem formulation

The concept of a balanced schedule presented in Section 1.1.3 is intuitively understood; however, it is not precise. For instance, it is not clear how to compare two schedules from Example 1.1: BDCBADBCDB and BDBCADCBDB. The part usage per hour is identical in both cases (the same 10 units are produced each hour), but the sequences are not identical. More precise measures of the schedule balance are proposed in the literature. We present them using the data from Example 1.1. Let us first we introduce the necessary notation.

Let us assume that we have a production system with L levels and n_l outputs at each level, $l = 1, \dots, L$. The demand for output i at level l is denoted by δ_{il}, $l = 1, \dots, L$, $i = 1, \dots, n_l$. For each output i, we denote by t_{ilp} the number of units of output i at level l required by the finished product p, $l = 1, \dots, L$, $i = 1, \dots, n_l$, $p = 1, \dots, n_1$. Obviously,

$$t_{i1p} = \begin{cases} 1 & \text{if } i = p, \\ 0 & \text{otherwise.} \end{cases}$$

Values t_{ilp} are obtained by performing the so-called explosion of the bill of material. The output demands for each finished product in Example 1.1 are presented in Table 1.1. Using the data from the BOM explosion and the finished product demand δ_{p1}, determined in the production plan, the total demand for output i at level l, $l = 2, \ldots, L$, $i = 1, \ldots, n_l$, is calculated as

$$\delta_{il} = \sum_{p=1}^{n_1} t_{ilp}\delta_{p1}. \tag{5.1}$$

The total demand D_l at level l is then calculated as $D_l = \sum_{i=1}^{n_l} \delta_{il}$. The total demand for finished products D_1 has to be met during the planning horizon. We say that there are D_1 consecutive stages and each product must be assigned to exactly one stage. Since preemption is not allowed, in order to construct the schedule, it is enough to define the sequence in which the finished products are assembled.

In Example 1.1 we considered 4 levels. The demand at level 1, $\delta i1$, $i = 1, \ldots, n_1$ is determined by the weekly production demand. The demands at lower levels are calculated using formula (5.1) and presented in Table 1.3. Although the total demand at level 1, D_1 is equal to 400, the balanced weekly schedule proposed in Chapter 1 is a concatenation of 40 identical sequences of 10 products each, so it is sufficient to provide a schedule for the first 10 stages.

Let us denote by x_{ilk} the number of units of output i scheduled at level l up to stage k (*cumulative production volume* of output i at level l in stage k), and by y_{lk}, $y_{lk} = \sum_{i=1}^{n_l} x_{ilk}$, the cumulative production volume of level l in stage k. We define the *production rate* of output i at level l in stage k, as the proportion of the cumulative production volume of output i at level l in stage k to the cumulative production volume at level l, x_{ilk}/y_{lk}. For organizational reasons, explained in Chapter 1, it is desired that the production rate be constant during the planning horizon. In order to meet the demand the production rate should be equal to an ideal production rate ρ_{il} of output i at level l, i.e., the ratio of the demand at level l created by output i, $\rho_{il} = \delta_{il}/D_l$. Moreover, we may consider weights w_{il} which characterize the relative importance of achieving the ideal production rate of output i at level l.

For simplicity, in Example 1.1 we assume $w_{il} = 1$, $l = 1, \ldots, L$, $i = 1, \ldots, n_l$. The ideal production rates of all outputs are calculated using the demand data from Tables 1.2 and 1.3. They are presented in column ρ_{il} in Table 5.1.

Since our goal is to keep constant production rate of each output throughout the planning horizon (i.e. equal in each stage k), the ideal cumulative production volume of output i at level l in stage k is the fraction of the cumulative production at level l determined by the ideal production rate of output i, and it is calculated as $\rho_{il} y_{lk}$. Now, for each i, l, k we can calculate the deviation of the cumulative production volume from the ideal one as: $|x_{ilk} - y_{lk}\rho_{il}|$. The relevant deviations calculated for schedule S_1 from Example 1.1 are given in Table 5.1.

The optimization criterion may be the minimization of the total absolute deviation calculated as

$$\sum_l \sum_i \sum_k |x_{ilk} - y_{lk}\rho_{il}|. \tag{5.2}$$

Let us calculate the total absolute deviation in schedule S_1 by simply adding the relevant deviations given in Table 5.1. Similarly, we can calculate the deviations in schedule S_2. We obtain 99.29 for S_1, and 120.46 for S_2.

Another optimization criterion is the minimization of the maximum deviation occurring for any output in any stage. The corresponding objective function, sometimes called a *bottleneck criterion*, is formulated as:

$$\max_{i,l,k} \{|x_{ilk} - y_{lk}\rho_{il}|\}. \tag{5.3}$$

In our example, the maximum absolute deviation equals 1.71 for schedule S_1, and 3.11 for schedule S_2. Although the sequences S_1 and S_2 look very similar, they give different values of both objective functions. With respect to both optimality criteria schedule S_1 is better balanced than S_2.

Functions (5.2) (total absolute deviation) and (5.3) (maximum absolute deviation) are most commonly examined in the literature ([188, 165, 87]), although other functions may be also used to calculate the cost of deviation from the ideal schedule. If the relative importance of particular outputs or levels differs significantly, then the minimization of the weighted maximum deviation (so-called *min-max problem*), defined by formula (5.4), or the total weighted deviation (so-called *min-sum problem*), defined by formula (5.5), may be an appropriate objective.

$$\max_{i,l,k} \{w_{il}|x_{ilk} - y_{lk}\rho_{il}|\} \tag{5.4}$$

$$\sum_l \sum_i \sum_k w_{il}|x_{ilk} - y_{lk}\rho_{il}| \tag{5.5}$$

Table 5.1. Deviations from ideal production in schedule S_1

	δ_{il}	D_l	ρ_{il}	1	2	3	4	5	6	7	8	9	10
Level 1 ($l = 1$)		400											
Product A	40		0.10	0.10	0.20	0.30	0.40	0.50	0.40	0.30	0.20	0.10	0
Product B	160		0.40	0.60	0.20	0.20	0.40	0	0.40	0.20	0.20	0.60	0
Product C	80		0.20	0.20	0.40	0.40	0.20	0	0.20	0.40	0.40	0.20	0
Product D	120		0.30	0.30	0.40	0.10	0.20	0.50	0.20	0.10	0.40	0.30	0
Level 2 ($l = 2$)		800											
Module X.1.a	160		0.20	0.40	0.20	0.20	0.60	0	0.60	0.20	0.20	0.40	0
Module X.1.b	240		0.60	0.20	0.20	0.60	0.00	0.60	0.20	0.20	0.40	0.00	0
Module X.2.a	120		0.70	0.60	0.10	0.20	0.50	0.20	0.10	0.60	0.30	0.00	0
Module X.2.b	280		0.30	0.60	0.10	0.20	0.50	0.20	0.10	0.60	0.30	0.00	0
Level 3 ($l = 3$)		3240											
Part X.1.1.a	200		0.062	0.43	0.07	0.54	0.98	0.47	0.98	0.54	0.07	0.43	0
Part X.1.1.b	400		0.123	0.14	0.85	0.91	1.05	0.06	1.05	0.91	0.85	0.14	0
Part X.1.2.a	400		0.123	0.14	1.15	0.09	0.05	1.06	0.05	0.09	1.15	0.14	0
Part X.1.2.b	200		0.062	0.43	0.93	0.46	0.02	0.47	0.02	0.46	0.93	0.43	0
Part X.2.1.a	440		0.136	1.05	0.04	1.40	0.35	1.43	0.35	1.40	0.04	1.05	0
Part X.2.1.b	520		0.160	1.12	0.59	0.99	0.14	1.58	0.14	0.99	0.59	1.12	0
Part X.2.2.a	120		0.037	0.26	0.56	0.93	1.19	1.48	1.19	0.93	0.56	0.26	0
Part X.2.2.b	720		0.222	1.44	1.67	0.56	0.89	1.11	0.89	0.56	1.67	1.44	0
Part X.2.2.c	240		0.074	0.52	1.11	1.15	0.63	0.04	0.63	1.15	1.11	0.52	0
Level 4 ($l = 4$)		2880											
Material 1	1400		0.486	1.60	1.71	0.21	0.90	1.50	0.90	0.21	1.71	1.11	0
Material 2	680		0.236	1.65	1.54	0.04	0.85	1.50	0.85	0.04	1.54	0.89	0
Material 3	800		0.278	0.06	0.17	0.17	0.06	0	0.06	0.17	0.17	0.22	0

Very often, instead of the absolute value, the quadratic deviation is applied to measure the deviation from the ideal production rate. Formula (5.6) expresses the total quadratic deviation.

$$\sum_{l=1}^{L}\sum_{i=1}^{n_l}\sum_{k=1}^{D_1}(x_{ilk} - y_{lk}\rho_{il})^2 \tag{5.6}$$

The values of the total quadratic deviation calculated for schedules S_1 and S_2 from Example 1.1 are 93.10, and 145.33, respectively. The total absolute deviation and the total quadratic deviation may produce very similar results. The difference is that the quadratic function penalizes large deviations more severely, so several small deviations may influence the value of the objective function (5.6) less than a single but large deviation.

In general, any l_p norm may be used to measure the cost of the deviation. If we denote by $f_{il}(x_{ilk} - y_{l\rho_{il}})$ the cost of deviation of the cumulative production volume of output i at level l in stage k from the ideal production volume, the total deviation cost is defined by formula (5.7) and the maximum deviation cost by formula (5.8).

$$\sum_{i=1}^{n_1}\sum_{l=1}^{L}\sum_{k=1}^{D_1} f_{il}(x_{ilk} - y_{lk}\rho_{il}) \tag{5.7}$$

$$\max_{i,l,k}\{f_{il}(x_{ilk} - y_{lk}\rho_{il})\} \tag{5.8}$$

Other optimality criteria are proposed in [188]. They include minimization of the following functions

$$\sum_{k=1}^{D_1}\sum_{l=1}^{L}\sum_{i=1}^{n_l} w_l\left(\frac{x_{ilk}}{ylk} - \rho_{il}\right)^2 \tag{5.9}$$

$$\sum_{k=1}^{D_1}\sum_{l=1}^{L}\sum_{i=1}^{n_l} w_l\left|\frac{x_{ilk}}{ylk} - \rho_{il}\right| \tag{5.10}$$

$$\sum_{k=1}^{D_1}\sum_{l=1}^{L}\sum_{i=1}^{n_l} w_l\left(x_{ilk} - \frac{\delta_{il}k}{D_1}\right)^2 \tag{5.11}$$

where the weights w_{il} of all outputs at the same level are equal, i.e. $w_{il} = w_l$, $i = 1, \ldots, n_l$.

According to Miltenburg, criterion (5.6) expresses the objective of schedule balancing best of all proposed criteria. First of all, as stated above, the quadratic function penalizes large deviations more severely, so it is more appropriate than the absolute deviation. Thus objective (5.9) is preferred to (5.10). The advantage of objective (5.6) over (5.9) follows from the fact that in the latter one deviations in the earlier stages contribute more to the objective function than in the later stages. Finally, the objective (5.6) is chosen over (5.11) because it assumes that the ideal cumulative production volume $\delta_{il}y_{lk}/D_l$ of an output is related to the production volume at the relevant level, not just to the production volume $\delta_{il}k/D_1$ at level 1. Notice that if we assume that the stage length is constant, then k may be a measure of time, so $\delta_{il}k/D_1$ represents the expected proportion of the production volume of output i at level l completed in the first k stages.

In order to complete the formulation of the problem, let us define the feasible region. Any of the objective functions defined above is minimized subject to the following constraints.

$$x_{ilk} = \sum_{p=1}^{n_1} t_{ilp}x_{p1k}, i = 1, \ldots, n_l, l = 1, \ldots, L, k = 1, \ldots, D_1, \quad (5.12)$$

$$y_{lk} = \sum_{i=1}^{n_l} x_{ilk}, l = 1, \ldots, L, k = 1, \ldots, D_1, \quad (5.13)$$

$$x_{p1D_1} = \delta_{p1}, x_{p10} = 0, p = 1, \ldots, n_1, \quad (5.14)$$

$$0 \leq x_{p1k} - x_{p1(k-1)} \leq 1, p = 1, \ldots, n_1, k = 1, \ldots, D_1, \quad (5.15)$$

$$k = \sum_{p=1}^{n_1} x_{p1k}, k = 1, \ldots, D_1, \quad (5.16)$$

$$x_{ilk} \in N, i = 1, \ldots, n_l, l = 1, \ldots, L, k = 1, \ldots, D_1. \quad (5.17)$$

Constraint (5.12) ensures that the total production volume of output i at level l by the end of stage k is determined explicitly by the quantity of products completed on the final assembly line (at level 1) by stage k. Constraint (5.13) establishes the relation between the cumulative production of each output and the total cumulative production at level l, through stages $1, \ldots, k$. Constraint (5.14) ensures that the demand for each product is met. Constraint (5.15) guarantees that the number of products completed on the assembly line (at level 1) by stage k is a non-decreasing function of the stage number k. Finally, constraints (5.15),

(5.16), and (5.17) guarantee that exactly one product is scheduled on the assembly line in each stage from 1 through D_1. The problem is to find an assignment of products to consecutive stages which minimizes the chosen criterion. The assignment is represented as matrix $\mathbf{X} = \{x_{p1k}\}$, $p = 1, \ldots, n_1$, $k = 1, \ldots, D_1$, of the cumulative production volume of each finished product at level 1.

In the following sections we discuss the multi-level mixed-model scheduling problems with criteria (5.2) and (5.3) in more detail.

5.1.2 Minimizing the maximum deviation

In this section we consider the multi-level just-in-time scheduling problem with the objective to minimize the maximum deviation from the ideal production rate. In general, the problem is to minimize the objective function defined by formula (5.8) under the constraints (5.12) to (5.17). In the literature, the most often examined cost function is the weighted absolute value of the deviation. The objective is then formulated as:

$$\text{minimize} \quad \max_{i,l,k} w_{il}|x_{ilk} - y_{lk}\rho_{il}| \tag{5.18}$$

where $w_{il} \geq 0$ is the weight representing the relative importance of meeting the ideal cumulative production volume of output i at level l.

Kubiak et al. prove in [165] that the multi-level just-in-time scheduling problem with two levels, identical weights $w_{il} = 1$, $l = 1, \ldots, L$, $i = 1, \ldots, n_l$, and the objective (5.18) is strongly NP-hard. The proof consists in constructing a pseudo-polynomial transformation from the 3-partition problem.

Below we present a dynamic programming procedure and two heuristic algorithms developed to solve the multi-level just-in-time scheduling problem with the objective (5.18).

Let us start with a transformation of the objective function. Considering the constraints defined by formulas (5.12), and (5.13), we can rewrite the objective function as follows.

$$\begin{aligned} w_{il}|x_{ilk} - y_{lk}\rho_{il}| &= |w_{il}(x_{ilk} - y_{lk}\rho_{il})| \\ &= \left| w_{il}\left(\sum_{p=1}^{n_1} t_{ilp}x_{p1k} - \rho_{il} \sum_{p=1}^{n_1} \sum_{h=1}^{n_l} t_{hlp}x_{p1k} \right) \right| \\ &= \left| w_{il} \sum_{p=1}^{n_1} \left(t_{ilp}x_{p1k} - \rho_{il} \sum_{h=1}^{n_l} t_{hlp}x_{p1k} \right) \right| \end{aligned}$$

$$= \left| \sum_{p=1}^{n_1} w_{il} \left(t_{ilp} - \rho_{il} \sum_{h=1}^{n_l} t_{hlp} \right) x_{p1k} \right|$$
$$= \left| \sum_{p=1}^{n_1} \gamma_{ilp} x_{p1k} \right|$$

where

$$\gamma_{ilp} = w_{il} \left(t_{ilp} - \rho_{il} \sum_{h=1}^{n_l} t_{hlp} \right).$$

Notice that the coefficients γ_{ilp} do not depend on the schedule and can be calculated in advance. Thus, it is clear that the value of the objective function depends only on the production sequence at the assembly level. We construct a n x n_1 matrix $\mathbf{\Gamma}$ (*gamma matrix*) of coefficients γ_{ilp}, where $n = \sum_{l=1}^{L} n_l$ is the total number of outputs at all levels. The rows of $\mathbf{\Gamma}$ correspond to particular outputs at the relevant production levels and the columns correspond to finished products, i.e. outputs at level 1. The cumulative production volume of product i, $i = 1, \ldots, n_1$, in stage k, $k = 1, \ldots, D_1$, is represented as a column vector $\mathbf{X}_k$, of n_1 entries x_{p1k}, $p = 1, \ldots, n_1$. Thus, vector $\mathbf{\Gamma X}_k$ represents the weighted deviations of all outputs i at level l in stage k. The *maximum norm* of a vector $\mathbf{X} = [x_1, \ldots, x_n]$, is defined as $||\mathbf{X}||_1 = \max_{1 \le i \le n} \{x_i\}$. Finally, the maximum absolute deviation in stage k is calculated using the following equation:

$$\max_{il} \{w_{il} |x_{ilk} - y_{lk}\rho_{il}|\} = ||\mathbf{\Gamma X}_k||_1. \tag{5.19}$$

The objective (5.18) can be now formulated as follows:

$$\text{minimize } \max_{k} ||\mathbf{\Gamma X}_k||_1. \tag{5.20}$$

Although the considered problem is NP-hard, an optimal schedule may be constructed using the dynamic programming approach. The dynamic programming algorithm presented below was proposed in [165]. Obviously, its computational complexity is exponential; however, it is much more efficient than the explicit enumeration.

The dynamic programming algorithm is defined as follows. Let $\delta = (\delta_{11}, \delta_{21}, \ldots, \delta_{n_1 1})$ be the vector of finished product demands. By $\boldsymbol{\epsilon}_i$, $i = 1, \ldots, n_1$, we denote a vector with all entries equal to zero except row i, where the entry equals 1.

In each stage k a decision is made which product is going to be scheduled in this stage. A state of the schedule in stage k is defined as

$\mathbf{X}_k = (x_{11k}, \ldots, x_{n_1 1k})$, where $x_{i1k} \leq \delta_{i1}$ is the cumulative production volume of product i at level 1 in stage k. We assume that $\mathbf{X}_0 \equiv 0$. The stage index k is then equal to the total cumulative production volume at level 1, i.e. $k = |\mathbf{X}_k| = \sum_{i=1}^{n_1} x_{i1k}$, and $||\mathbf{\Gamma X}_k||_1$ is the maximum deviation of the actual production over all outputs (parts and products) in state $\mathbf{X}_k$. The decision made in state $\mathbf{X}_k$, that product i should be scheduled in stage k, leads to a new state $\mathbf{X}_{k+1} = \mathbf{X}_k + \boldsymbol{\epsilon}_i$. Thus, $||\mathbf{\Gamma X}_{k+1}||_1$ depends only on $\mathbf{X}_k$ and $\boldsymbol{\epsilon}_i$. Let us denote by $\Phi(\mathbf{X}_k, \boldsymbol{\epsilon}_i)$ the minimum value of the maximum deviation by stage k, if product i is scheduled in stage k, i.e. $\mathbf{X}_{k-1} = \mathbf{X}_k - \boldsymbol{\epsilon}_i$, and by $\Phi^*(\mathbf{X}_k)$ the minimum of $\Phi(\mathbf{X}_k, \boldsymbol{\epsilon}_i)$ over all decisions possible in state $\mathbf{X}_k$,

$$\Phi^*(\mathbf{X}_k) = \min_i \{\Phi(\mathbf{X}_k, \boldsymbol{\epsilon}_i)\}.$$

We calculate $\Phi(\mathbf{X}_k)$ recursively, using the following equation:

$$\Phi(\mathbf{X}_k, \boldsymbol{\epsilon}_i) = \max\{\Phi^*(\mathbf{X}_k - \boldsymbol{\epsilon}_i), ||\mathbf{\Gamma X}_k||_1\}.$$

Concluding, the following recursive relation holds for $\Phi(\mathbf{X}_k)$:

$$\Phi(\emptyset) = \Phi^*(\mathbf{X}_0) = 0,$$

$$\Phi^*(\mathbf{X}_k) = \min\{\max\{\Phi^*(\mathbf{X}_k - \boldsymbol{\epsilon}_i), ||\mathbf{\Gamma X}_k||_1\} : i = 1, \ldots, n_1, x_{i1k} - 1 \geq 0\}.$$

Obviously, $\Phi^*(\mathbf{X}_k) \geq 0$ for any state $\mathbf{X}_k$ and $||\mathbf{\Gamma d}||_1 = 0$. The number of states may grow exponentially with n_1, but its growth rate is polynomial in D_1. Thus, the dynamic programming algorithm may be effective if the number of different products is small even if the total demand D_1 is large.

Although the complexity of the procedure is $O(n_1 n \prod_{i+1}^{n_1} (\delta_i + 1))$, it is much lower than the number of all feasible schedules, which equals $D_1!/(\delta_1!\delta_2!, \ldots \delta_{n_1}!)$. Moreover, in order to improve the performance of the dynamic programming algorithm, a filtering method is proposed in [165]. The idea is to find a good heuristic solution of the problem and eliminate states (and their successors) that would lead to solutions worse that the heuristic ones. The filtering method is quite easy to implement and may be very effective, especially if large deviations appear in early stages of the schedule. The authors report solving optimally problems with $n_1 = 12$ and $D_1 = 500$ in a 4-level system. They used the following two heuristic algorithms to calculate the filter value.

Algorithm 5.1 (One-stage heuristic [165]).

1. Set $\mathbf{X}_0 \equiv 0$, $k = 1$.

2. Schedule in stage k product p with minimum value of $||\mathbf{\Gamma}(\mathbf{X}_k+\boldsymbol{\epsilon}_p)||_1$.
3. If $k < D_1$ then set $\mathbf{X}_{k+1} = \mathbf{X}_k + \boldsymbol{\epsilon}_p$ and $k = k + 1$ and go to step 2, else stop.

The one-stage heuristic is a greedy algorithm, scheduling in each stage the output that minimizes the objective function from the point of view of the current state. In the following two-stage heuristic, in each stage we schedule the product which minimizes the objective function taking into account the best possible choice in the next stage.

Algorithm 5.2 (Two-stage heuristic [165]).

1. Set $k = 1$.
2. Schedule in stage k product p with minimum value of

$$\max\{||\mathbf{\Gamma}(\mathbf{X}_k + \boldsymbol{\epsilon}_p)||_1, \min_q ||\mathbf{\Gamma}(\mathbf{X}_k + \boldsymbol{\epsilon}_p + \boldsymbol{\epsilon}_q)||_1\}.$$

3. If $k < D_1$ then set $\mathbf{X}_{k+1} = \mathbf{X}_k + \boldsymbol{\epsilon}_p$ and $k = k + 1$ and go to step 2, else stop.

Since the running time of both heuristics is negligible, in the experiments reported in [165] both heuristics were run and the smaller value of the objective function was chosen to serve as the filter.

It is very likely that an efficient branch and bound algorithm can be constructed for the ORV problem, since the heuristics provide good upper bounds for the maximum absolute deviation.

Finally, let us formulate so-called *pegging* assumption. According to the APICS dictionary [76], a *pegged requirement* is a requirement that shows next-level parent item (or customer order) as the source of the demand. It is applied in MRP systems to identify the destination of each output. The pegging information can be used to go up through the MRP records from a raw material gross requirement to some future customer order. Pegging is sometimes treated as where-used data. Identification of the source of demand is especially important in manufacturing high quality products, where it is essential to know which parts were placed into which product. If a damaged item should be replaced, it is easy to obtain information which finished products are affected by defective parts and to decrease the costs of replacement. Such situations sometimes occur, for example, in the automotive industry. The first mathematical formulation of pegging in the JIT environment was proposed by Goldstein and Miltenburg in [106]. In [223] Steiner and Yeomans show that the multi-level just-in-time scheduling problem with pegging reduces to a single-level problem. Below we formulate the multi-level problem with pegging and show the reduction to the single-level problem.

Under the pegging assumption, outputs at levels $2, \ldots, L$ are dedicated (or *pegged*) to the particular finished products into which they are assembled. In other words, we can assume that outputs at lower levels create disjoint sets, depending on which product they are pegged to. In consequence, the values t_{ilp} and t_{ilq} have to be distinguished for $p \neq q$. Thus, the ideal cumulative production volume of output i at level l pegged to product p in stage k is now calculated as $kt_{ilp}\rho_{p1}$, and is independent of the final assembly sequence. As a result, the objective is formulated as follows:

$$\text{minimize} \quad \max_{i,l,p,k} \{w_{p1}|x_{p1k} - k\rho_{p1}| + w_{il}|t_{ilp}x_{p1k} - kt_{ilp}\rho_{p1}|\} \tag{5.21}$$

where $l = 2, \ldots L$, $k = 1, \ldots, D_1$, $p = 1, \ldots, n_1$, and $i = 1, \ldots, n_l$.

Let us observe that $y_{1k} = k$, and $t_{ilp} \geq 0$. Taking into account the relation defined by equation (5.1), the objective may be rewritten in the following way:

$$\text{minimize} \quad \max_{l,p,k} \{W_p|x_{p1k} - y_{1k}\rho_{p1}|\} \tag{5.22}$$

where $W_p = \max_{il} \{w_{il}t_{ilp}\}$.

Now we can see that the level index is superfluous, so it can be dropped. We finally obtain:

$$\text{minimize} \quad \max_{p,k} \{W_p|x_{pk} - y_k\rho_p|\} \tag{5.23}$$

where $p = 1, \ldots, n$, and $k = 1, \ldots, D$.

This is the single-level just-in-time scheduling problem which is formulated and discussed in Section 5.2.

5.1.3 Minimizing the total deviation

Another important optimality criterion considered in multi-level production systems is the total cost of the output rate variation. In general, the problem is defined as minimization of the objective function given by formula (5.7) under the constraints (5.12) to (5.17). In the literature, the quadratic function is the most often examined cost function.

The problem was first formulated by Miltenburg and Sinnamon in [188], where some heuristic algorithms for solving the problem with objective function defined by formula (5.6) were also proposed. Some improvements of the Goal Chasing Method proposed by Monden in [191] and preliminary computational analysis of the heuristics developed for the two-level ORV problem are presented in [23].

In [153] the multi-level problem with the objective function defined by formula (5.6) is proved to be NP-hard already for only two levels ($L = 2$), two outputs at level 2, and identical weights. The proof is based on reduction from the SASJ (Scheduling Around the Shortest Job) problem which is NP-hard in the ordinary sense (see [154]).

Let us observe that the objective function defined by equation (5.6) can be presented in a matrix form. From the previous section we have:

$$w_{il}(x_{ilk} - y_{lk}\rho_{il})^2 = \sum_{p=1}^{n_1}\left(\sqrt{w_{il}}t_{ilp} - \rho_{il}\sum_{h=1}^{n_1} t_{hlp}x_{p1k}\right)^2 \tag{5.24}$$

Let us set

$$\omega_{ilp} = \sqrt{w_{il}}\left(t_{ilp} - \rho_{il}\sum_{h=1}^{n_1} t_{hlp}\right)$$

and construct $\mathbf{\Omega}$ as an n x n_1 matrix, $n = \sum_{l=1}^{L} n_l$, with entries ω_{ilp}. Recall that the Euclidean norm of vector $\mathbf{X}$ is calculated as $||\mathbf{X}||_2 = \sqrt{\sum_{i=1}^{n} x_{ij}^2}$. We obtain

$$\begin{aligned}
&\sum_{l=1}^{L}\sum_{i=1}^{n_l}(x_{ilk} - y_{lk}\rho_{il})^2 \\
&= \sum_{l=1}^{L}\sum_{i=1}^{n_l}\left(\sum_{p=1}^{n_1}\sqrt{(w_{il})}(t_{ilp} - \rho_{il}\sum_{h=1}^{n_1} t_{hlp})x_{p1k}\right)^2 \\
&= \sum_{l=1}^{L}\sum_{i=1}^{n_l}\left(\sum_{p=1}^{n_1}\omega_{ilp}x_{p1k}\right)^2 \\
&= (||\mathbf{\Omega}\mathbf{X}_k||_2)^2.
\end{aligned}$$

where vector $\mathbf{X}_k$ is defined as in the previous section. The objective to minimize the total weighted quadratic deviation may be now formulated as:

$$\text{minimize} \sum_{k=1}^{D_1}(||\Omega\mathbf{X}_k||_2)^2 \tag{5.25}$$

Analogously to the problem with minimization of the maximum deviation, a dynamic programming procedure was proposed in [165] for minimization of the total deviation. The corresponding recursive relation is defined as follows:

$$\Phi(\emptyset) = \Phi^*(\mathbf{X}_0) = 0,$$

$$\Phi^*(\mathbf{X}_k) = \min\{\Phi^*(\mathbf{X}_k - e_i) + ||\Omega \mathbf{X}_k||_2)^2\} : i = 1, \ldots, n_1, x_i - 1 \geq 0\}.$$

The function $\Phi(\mathbf{X}_k) \geq 0$, $k = 1, \ldots, D_1$, and $||\Omega \mathbf{d}||_2)^2 = 0$. Simple heuristics, analogous to Algorithms 5.1 and 5.2, may be used as filters in order to reduce the running time of the dynamic programming algorithm.

In [188] Miltenburg and Sinnamon propose heuristic algorithms for minimization of the total quadratic deviation and analyze the effect of the weights. Below, we briefly present the heuristics.

First, the algorithm assigns products to consecutive stages, using a priority rule based on the deviation generated by each product p at the first level. Next, in each stage the product with the smallest priority is scheduled. We define the following coefficients, used in the algorithm to calculate the priorities of products,

$$\beta_{plk} = \sum_{j=1}^{n_l} w_{pl} \left[(x_{jl(k-1)} + t_{jlp}) - \left(\sum_{p=1}^{n_l} \alpha_{lp} \right) \rho_{jl} \right]^2,$$

where

$$\alpha_{lp} = \sum_{h=1}^{n_l} t_{hlp}.$$

Following is a more formal presentation of the algorithm.

Algorithm 5.3 (Miltenburg single-stage heuristic [188]).

1. Set $k = 1$.
2. Schedule in stage k product p for which the following value is minimal
$$w_{p1}(x_{11(k-1)} - k\rho_{p1}) + 0.5 \sum_{l=2}^{L} \beta_{plk}$$
3. If $k < D_1$ then set $k = k + 1$ and go to step 2, else stop.

In the original formulation of the algorithm, only weights w_l equal for all outputs at level l were considered, i.e. $w_{pl} = w_l$, for $p = 1, \ldots, n_l$. However, the generalization presented above is quite natural. The computational complexity of Algorithm 5.3 is $O(D_1(n_1(\sum_{l=2}^{L} n_l)))$.

Computational experiments show that in the schedules obtained by Algorithm 5.3 the deviations at levels 3 and 4 almost always exceed the deviations at levels 1 and 2. It is also the case in our example (see Table 5.2). This may be easily explained by the greedy nature of Algorithm 5.3. Miltenburg points out that it is easy to control the deviations at particular levels by using appropriate weights. The most popular approach is to set the weights at the most important level at 1, and all the other weights at 0.

Table 5.2. Total quadratic deviations at particular levels

Level	Quadratic deviations	
l	schedule S_1	schedule S_2
1	3.70	5.30
2	5.50	8.10
3	58.18	81.55
4	25.72	50.38
Total	93.10	145.33

The second heuristic is slightly more complex. It also constructs the schedule stage by stage. In each stage k, for each candidate product i to be scheduled in this stage, the variation at level 1 is calculated, assuming that in stage $k+1$ product j minimizing the value $w_{j1}(x_{11(k)} - (k+1)\rho_{j1}) + 0.5 \sum_{l=2}^{L} \beta_{jl(k+1)}$ is chosen. The product that minimizes the variation is scheduled in stage k. Following is a more formal presentation of the two-stage heuristic.

Algorithm 5.4 (Miltenburg two-stage heuristic [188]).

1. Set $k = 1$.
2. Set $p = 1$.
3. Tentatively schedule product p in stage k and calculate the corresponding variation

$$V1_p = \sum_{l=1}^{L} \sum_{p=1}^{n_l} w_{pl}(x_{plk} - y_{lk}\rho_{pl})$$

.

4. Find product j with the lowest value of the formula

$$w_{j1}(x_{11(k)} - (k+1)\rho_{j1}) + 0.5\sum_{l=2}^{L}\beta_{jl(k+1)}$$

   and calculate the variation $V2_p$ in stage $k+1$; set $V_p = V1_p + V2_p$.
5. If $p < n_1$ then set $p = p+1$ and go to step 3, else go to step 6.
6. Schedule product p with the lowest V_p in stage k.
7. If $k < D_1$ then set $k = k+1$ and go to step 2, else stop.

The complexity of Algorithm 5.4 is $O(n_1^2(\sum_{l=2}^{L} n_l)^2)$ and is obviously higher than the complexity of Algorithm 5.3.

The third heuristic approach to solving the multi-level just-in-time scheduling problem with minimization of total quadratic deviation is the Goal Chasing Method, GCM, developed and used in Toyota to schedule automobiles on the final assembly line. The algorithm dedicated to a two level system was presented by Monden [191].

Algorithm 5.5 (Goal Chasing Method - GCM [191, 188]).

1. Set $k = 1$.
2. Schedule in stage k product p for which the following value is minimal

$$GCM_{pk} = \sum_{i=1}^{n_2}[(x_{i2(k-1)} + t_{i2p}) - k\delta_{i2}/D_1]^2.$$

3. If $k < D_1$ then set $k = k+1$ and go to step 2, else stop.

In [188] Miltenburg and Sinnamon propose an extension of the GCM to four levels. This extension in turn, may be easily generalized for an arbitrary number of levels L as follows. Let us denote

$$\beta_{plk} = \sum_{i=1}^{n_l}[(x_{il(k-1)} + t_{ilp}) - k\delta_{il}/D_1]^2.$$

Algorithm 5.6 (Extended Goal Chasing Method - EGCM [188]).

1. Set $k = 1$.
2. Schedule in stage k product p for which the following value is minimal

$$EGCM_{pk} = \sum_{l=1}^{L}\beta_{plk}$$

3. If $k < D_1$ then set $k = k+1$ and go to step 2, else stop.

Although the rule used in EGCM is similar to the rule used in Algorithm 5.3, it is easy to notice that EGCM aims at minimizing a different objective function, namely the function defined by formula (5.26).

$$\sum_{k=1}^{D_1}\sum_{l=1}^{L}\sum_{i=1}^{n_l} w_{il}(x_{ilk} - \delta_{il}k/D_1)^2 \tag{5.26}$$

The EGCM becomes identical with Algorithm 5.3 if we assume that $k/D_1 \approx (\sum_{i=1}^{n_l} x_{ilk})/D_j$, for any k, $k = 1,\dots,D_1$. This means that for all stages the actual production rate at level l is the same as the proportion of the time elapsed from the start of the schedule (if the stage length is constant). The EGCM rule may be also used as the decision rule in Algorithm 5.4.

Miltenburg and Sinnamon [188] present an example where each of the four heuristics (Algorithms 5.3, 5.4, 5.5, and 5.6) produces a different schedule.

Bautista et al. in [23] transform the problem considered by Monden to the problem of finding a minimum path in a directed acyclic graph. The graph consists of $D_1 + 1$ levels. A node at level k, $k = 1,\dots,D_1$, corresponds to vector $\mathbf{X}_k$ and is denoted by $X(k,h)$. The number of nodes depends on the level. Levels 0 and D_1 consist of single nodes: $X(0)$ corresponding to vector $\mathbf{X}_0 \equiv \mathbf{0}$, and $X(D_1)$, corresponding to vector $\mathbf{X}_{D_1} = [\delta_1,\dots,\delta_{n_1}]$, respectively. Only nodes at consecutive levels are connected. If vector $\mathbf{X}_{k+1,h} = \mathbf{X}_{k,h'} + \boldsymbol{\epsilon}_j$, for any j, $j = 1,\dots,n$, then nodes $X(k,h)$ and $X(k+1,h')$ are connected. Recall that $\boldsymbol{\epsilon}_j$ is the column vector with the j-th entry equal to one and all remaining entries equal to zero. Finally, with each node we associate a weight $A(k,h)$ defined by equation (5.27).

$$A(k,h) = \sum_{i=1}^{n_2}(\sum_{p=1}^{n_1} t_{i2p}x_{p1k} - k\rho_{i2})^2. \tag{5.27}$$

The problem of finding a schedule minimizing function (5.9) is now equivalent to finding the shortest path from node $X(0)$ to node $X(D_1)$ in the graph defined above. However, solving the shortest path problem may be time consuming if the number of nodes is large. The Goal Chasing Method may be used as a heuristic to build a path in the graph. Bautista et al. in [23] call this algorithm the Revised Goal Chasing Method. Let us denote by $succ(X(k,h))$ all successors of node $X(k,h)$ at level $k+1$.

Algorithm 5.7 (Revised Goal Chasing Method - RGCM). [23]

1. Set $k = 1$.
2. Let $X(k+1, h')$ be the node for which $A(k+1, h') = \min\{A(k+1, h) : X(k+1, h) \in succ(X(k, h))\}$. Set $X(k+1, h) = X(k+1, h')$.
3. If $k < D_1$ then set $k = k + 1$ and go to step 2, else stop.

Since the weights $A(k, h)$ can be calculated recursively, the revised Goal Chasing Method is very fast.

The three following procedures are proposed in [23] to improve the Goal Chasing Method:

- *Symmetry.* The idea is to create a path by concatenating two segments: one starting in node $X(0)$ and the second one starting in node $X(D)$. Arcs are added to the alternate segments.
- *Horizon.* This improvement implements the concept of the two-stage heuristic (Algorithm 5.4). For each node, all two-arc segments are evaluated and the first arc from the segment yielding the minimum deviation is chosen. Although more that two steps ahead could be considered, it is optimal from computational reasons to consider only two consecutive arcs.
- *Rate-preserving.* Here the goal is to have a mechanism which eliminates too long postponement of scheduling a product, which may result in large deviations in later stages. This goal is achieved by comparing the demand vectors of the initial state and the state remaining after stage k. All candidate nodes with deviation less that 5% from the best one are considered tied and the tie is resolved by choosing the node which results in the remaining demand vector closest to the initial one.

Computational experiments reported by Bautista et al. in [23] show that the best results are obtained by combining the two-step horizon with the rate-preserving mechanism.

Notice that finding a suboptimal solution with the proposed heuristic may require large computational effort, especially for large D_1, and thus solving real life problems becomes too expensive. Fortunately, quite often an optimal sequence is cyclic and the algorithm may stop after the first stage k for which the total deviation equals zero. The complete schedule is then obtained as a concatenation of D_1/k subsequences of length k, as Miltenburg and Sinnamon show in [188].

The multi-level problem with the objective to minimize the total deviation may be also considered under the pegging assumption (see [106, 223]). It is also reducible to a single-level problem with the objective function obtained as follows. Notice that the ideal cumulative production volume of output i at level l pegged to product p in stage k equals $kt_{ilp}\rho_{p1}$. Thus the objective function can be written as

$$\sum_{k=1}^{D_1}\sum_{p=1}^{n_1}\left(w_{p1}(x_{p1k}-k\rho_{p1})^2+\sum_{l=2}^{L}\sum_{i=1}^{n_l}w_{il}(t_{ilp}x_{p1k}-kt_{ilp}\rho_{p1})^2\right). \quad (5.28)$$

Taking into account relation (5.1), we obtain

$$\begin{aligned}&\sum_{k=1}^{D_1}\sum_{p=1}^{n_1}\sum_{l=1}^{L}\sum_{i=1}^{n_l}w_{il}t_{ilp}^2(x_{p1k}-k\rho_{p1})^2\\ &=\sum_{k=1}^{D_1}\sum_{p=1}^{n_1}(x_{p1k}-k\rho_{p1})^2\sum_{l=1}^{L}\sum_{i=1}^{n_l}w_{il}t_{ilp}^2.\end{aligned}$$

Let us denote

$$V_p=\sum_{l=1}^{L}\sum_{i=1}^{n_l}w_{il}t_{ilp}^2.$$

We can rewrite the objective function as

$$\sum_{k=1}^{D_1}V_p(x_{p1k}-k\rho_{p1})^2. \quad (5.29)$$

Now, we can drop the superfluous level subscript and substitute $k=y_k$ to obtain the following objective.

$$\text{minimize} \quad \sum_{k=1}^{D}V_p(x_{pk}-y_k\rho_p)^2. \quad (5.30)$$

The single-level problem will be considered in more detail in the next section.

5.2 The single-level scheduling problem

The motivation for examining the single-level JIT scheduling problem is twofold. First, it often happens that products require approximately the same number and mix of parts. In such case a constant rate of usage of each output may be achieved by considering only the demand rates of the finished products. The problem reduces to scheduling products at a single-level, i.e. level 1. Second, the multi-level problem with pegging also reduces to the single-level problem. The single-level problem is formulated by Miltenburg in [186], and is later called by Kubiak the *Product Rate Variation Problem* (PRV) in [153]. The optimality criteria considered are the same as in the general multi-level formulation. However, the constraints are significantly simplified and the number of variables is reduced.

In this section we formulate the PRV problem and present algorithms for solving the problem with the most important objectives: maximum and total deviation from the ideal cumulative production volume. Further, we examine a class of schedules for the PRV problem, called cyclic schedules. Finally, we present the transformation from the PRV problem to the problem of apportionment and examine the properties of the algorithms developed for solving the PRV problem.

5.2.1 Problem formulation

Assuming that a balanced schedule at level 1 guarantees low production rate variation at the remaining levels, the demands for finished products are the only input data. Let us consider a set of n products with production demands equal to $\delta_1, \delta_2, \ldots, \delta_n$ units, respectively, and the total demand $D = \sum_{i=1}^{n} \delta_i$. Let us denote by x_{ik} the cumulative production volume of product i in stage k. Notice that in the single-level system exactly k products are completed by stage k, so $y_k = k$. We define the ideal production rate of product i as the ratio of the total demand generated by product i, $\rho_i = \delta_i / D$. Thus the ideal cumulative production volume of output i in stage k is the fraction of the cumulative production determined by the ideal production rate of output i, and is calculated as $k\rho_i$. Now, for each i and k we can calculate the deviation of the cumulative production volume from the ideal volume as: $|x_{ik} - k\rho_{il}|$. The corresponding deviations at level 1 in Example 1.1 can be found in Table 5.1.

The objective functions considered are the total and maximum deviation cost, where the deviation cost can be given by any non-decreasing function. The functions most commonly used to evaluate the cost of deviation are the absolute value and the quadratic function.

In general, the maximum deviation cost is defined as follows:

$$\text{minimize} \qquad \max_{ik}\{f_i(x_{ik} - k\rho_i)\}. \tag{5.31}$$

This objective was introduced by Steiner and Yeomans [221] for $f_i(x_{ik} - k\rho_i) = |x_{ik} - k\rho_i|$. The maximum absolute deviations for schedules S_1 and S_2 from Example 1.1 are equal to 0.6 and 0.8, respectively.

The second objective is to minimize the total deviation cost, defined as:

$$\text{minimize} \qquad \sum_{i=1}^{n}\sum_{k=1}^{D} f_i(x_{ik} - k\rho_i) \tag{5.32}$$

Finally, the feasible region is determined by the following constraints:

$$\sum_{i=1}^{n} x_{ik} = k, \qquad k = 1, \ldots, D \tag{5.33}$$

$$0 \leq x_{ik+1} - x_{ik} \leq 1 \qquad n = 1, \ldots, n; k = 1, \ldots, D-1 \tag{5.34}$$

$$x_{iD} = \delta_i \qquad n = 1, \ldots, n \tag{5.35}$$

Constraint (5.33) guarantees that the cumulative production of all products completed by time unit k equals k. Recalling that the completion of each product requires one time unit, we can conclude that no idle time occurs in the schedule. Fulfilling constraint (5.34) is necessary to have exactly one product assigned to each time unit. Finally, constraint (5.35) assures that the demand for each product is satisfied.

5.2.2 Minimizing the maximum deviation

Let us first consider the PRV problem with the objective to minimize the maximum absolute deviation from the ideal production volume. The objective is defined as follows:

$$\text{minimize} \qquad \max_{ik} |k\rho_i - x_{ik}|. \tag{5.36}$$

Tijdeman [238] proves that a schedule with maximum deviation less than 1 can be found for each instance of the PRV problem. Tijdeman [238] considers the problem called the *chairman assignment problem*, formulated as follows.

There are n states with weights $\lambda_1, \ldots, \lambda_n$, $\sum_{i=1}^{n} \lambda_i = 1$, which form a union. Every year a chairman of the union is appointed in such a way that at any time the accumulated number of chairmen from state i is proportional to the weight λ_i. Let x_{ik} denote the number of chairmen representing state i in the first k years. The problem is to minimize the maximum deviation from an optimal chairmen distribution.

$$\text{minimize} \qquad \max_{ik} |k\lambda_i - x_{ik}| \tag{5.37}$$

It is easy to notice that if we substitute the weights λ_i by the relative demands of products ρ_i, $i = 1, \ldots, n$, we obtain the PRV problem (5.36). Meijer in [184] proved the following theorem, cited by Tijdeman in [238].

Theorem 5.8 ([184]). *Let λ_{ik} be a double sequence of non-negative numbers such that $\sum_{i=1}^{n} \lambda_{ik} = 1$ for $k = 1, \ldots$. For an infinite sequence S in $\{1, \ldots, n\}$ let x_{iv} be the number of i's in the v-prefix of S. Then there exists a sequence S in $\{1, \ldots, n\}$ such that*

$$\max_{iv} \left| \sum_{k=1}^{v} \lambda_{ik} - x_{iv} \right| \leq 1 - \frac{1}{2(n-1)}.$$

Theorem 5.24 addresses a more general case than the PRV problem, that is, a problem where the relative demand for particular products may vary over time. In the PRV formulation in Section 5.2.1 the relative demand ρ_i for product i is constant, so if we assign $\lambda_{ik} = \rho_i = \delta_i / D$, for $k = 1, \ldots$, then by Theorem 5.24 there exists an infinite sequence S such that

$$\max_{ik} |k\rho_i - x_{ik}| \leq 1 - \frac{1}{2(n-1)}. \tag{5.38}$$

A sequence satisfying (5.38) is constructed by Algorithm 5.9. Let $J_k, k = 1, \ldots, D$, be a set of products satisfying the following conditions in stage k:

$$k\rho_i - x_{ik-1} \geq \frac{1}{2n-2}, \tag{5.39}$$

and

$$\sigma_{ik} = \frac{1 - \frac{1}{2n-2} - (k\rho_i - x_{ik-1})}{\rho_i}, \tag{5.40}$$

where x_{ik-1} is the number of units of product i scheduled from stage 1 through $k - 1$.

Algorithm 5.9 (Tijdeman [238]).

1. Set $k = 1$ and $x_{i0} = 0$, for $i = 1, \ldots, n$.
2. Schedule in stage k product i from set J_k for which the value σ_{ik} is minimal.
3. If $k < D$ then set $k = k + 1$ and go to step 2, else stop.

The following example illustrates the Tijdeman algorithm.

Example 5.10. Let us consider 4 products with the following demands: $\delta_1 = 1, \delta_2 = 4, \delta_3 = 2$, and $\delta_4 = 3$. We obtain $D = 10$ and $\rho_1 = 0.1, \rho_2 = 0.4, \rho_3 = 0.2$, and $\rho_4 = 0.3$. Moreover, $1/(2n-1) = 0.143$. The values x_{ik}, and σ_{ik} are presented in Table 5.3. The values in boldface in column $(k\rho_i - x_{ik})$ indicate which products belong to set J_k, and the values in column σ_{ik}, indicate the product scheduled in stage k.

Table 5.3. Tijdeman algorithm for Example 5.10

k	x_{ik-1}	$k\rho_i - x_{ik}$				σ_{ik}			
1	0 0 0 0	0.1	**0.4**	**0.2**	**0.3**	–	**1.08**	3.17	1.78
2	0 1 0 0	**0.2**	-0.2	**0.4**	**0.6**	6.33	–	2.17	**0.78**
3	0 1 0 1	**0.3**	**0.2**	**0.6**	-0.1	5.33	1.58	**1.17**	–
4	0 1 1 1	**0.4**	**0.6**	-0.2	**0.2**	4.33	**0.58**	–	2.11
5	0 2 1 1	**0.5**	0	0	**0.5**	3.33	–	–	**1.11**
6	0 2 1 2	**0.6**	**0.4**	**0.2**	-0.2	2.33	**1.08**	3.17	–
7	0 3 1 2	**0.7**	-0.2	**0.4**	0.1	**1.33**	–	2.17	–
8	1 3 1 2	-0.2	**0.2**	**0.6**	**0.4**	–	1.58	**1.17**	1.44
9	1 3 2 2	-0.1	**0.6**	-0.2	**0.7**	–	0.58	–	**0.44**
10	1 3 2 3	0	**1**	0	0	–	**-0.42**	–	–
11	1 4 2 3	–	–	–	–	–	–	–	–

The schedule generated by the algorithm is determined by the sequence 2432421342. The Tijdeman algorithm constructs a schedule with $|k\rho_i - xik| < 1 - 1/(2n-1)$; however, schedules with a smaller maximum deviation may exist. Steiner and Yeomans [221] proposed an algorithm that finds an optimal schedule. They formulate the single-level scheduling problem considered by Miltenburg in [186] as the following decision problem. For a given vector of demands $[\delta_1, \ldots, \delta_n]$ and a constant F, does there exist a schedule with $max|x_{ik} - k\rho_i| \leq F$? An algorithm that answers this question in $O(D)$ time is based on the calculation of a time window for each unit j of each product i, such that if the product is completed within this time window, then the inequality

$$|j - k\rho_i| \leq F \tag{5.41}$$

holds. In order to find the time window for unit j of product i, Steiner and Yeomans analyze the absolute deviation from $k\rho_i$ for each unit j of product i as a function of its completion time.

Example 5.11. Let us consider a product with $\delta = 4$ and $D = 10$. Obviously, $\rho = 0.4$. A collection of graphs representing the functions $|j - k\rho_i|$ as functions of k, for each unit j, $j = 1, 2, 3, 4$, is presented in Figure 5.1. The intervals in which each unit of the product has to be completed are outlined below the graph.

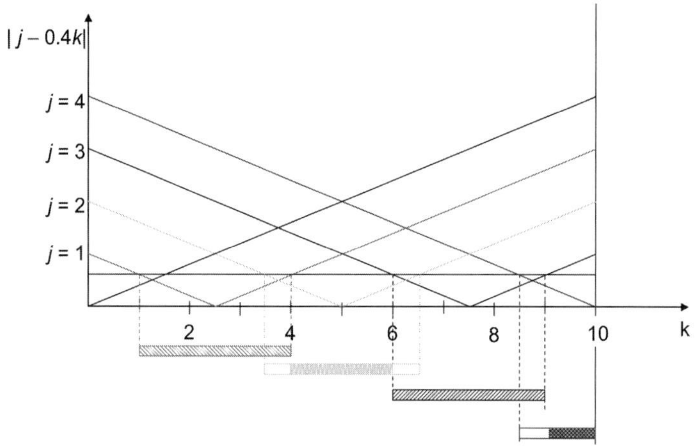

Fig. 5.1. Graphs of cost functions from Example 5.11.

It is easy to see from the graph that in order to satisfy inequality (5.41) unit j of product i may not be completed before time $E(i, j)$ defined as follows:

$$j - E(i,j)\rho_i \leq F < j - (E(i,j) - 1)\rho_i \tag{5.42}$$

Similarly, the latest completion time $L(i, j)$ for unit j of product i must satisfy the following inequality:

$$L(i,j)\rho_i - j \leq F < (L(i,j) + 1)\rho_i - j \tag{5.43}$$

Concluding, a solution to the PRV problem with $\max_{ik} |x_{ik} - k\rho_i| \leq F$ exists if and only if there exists a sequence that allocates the j-th unit of product i in the interval $[E(i, j), L(i, j)]$, where

$$E(i,j) = \left\lceil \frac{1}{\rho_i}(j - F) \right\rceil, \tag{5.44}$$

$$L(i,j) = \left\lfloor \frac{1}{\rho_i}(j + F) \right\rfloor. \tag{5.45}$$

Steiner and Yeomans prove that the existence of such sequence may be checked in $O(D)$ time by applying the Earliest Due Date algorithm. The Steiner-Yeomans algorithm calculates the values $E(i, j)$ and $L(i, j)$ for each pair (i, j), $i = 1, \ldots, n; j = 1, \ldots, \delta_i$, and assigns positions from $k = 1$ to $k = D$, each time selecting an unassigned pair (i, j) with the lowest value $L(i, j)$, from the set of all pairs satisfying the condition $E(i, j) \leq k \leq L(i, j)$. If in any stage k no pair (i, j) such that $E(i, j) \leq k \leq L(i, j)$ exists, then the conclusion is that a schedule with $\max_{ik} |x_{ik} - k\rho_i| \leq F$ does not exist. The algorithm is presented below in a more formal way.

Algorithm 5.12 (Steiner-Yeomans [221]).

1. Set $k = 1$ and $x_{i0} = 0$, $i = 1, \ldots, n$.
2. Set $i = 1$, $MIN = D$, and $j = 0$.
3. If $\left\lceil \frac{1}{\rho_i}(x_{ik-1} + 1 - F) \right\rceil \leq k \leq \left\lfloor \frac{1}{\rho_i}(x_{ik-1} + F) + 1 \right\rfloor$ and $x_{ik-1} < \delta_i$ then go to step 4, else go to step 5.
4. If $\left\lfloor \frac{1}{\rho_i}(x_{ik-1} + F) + 1 \right\rfloor < MIN$ then set $MIN = \left\lfloor \frac{1}{\rho_i}(x_{ik-1} + F) + 1 \right\rfloor$ and set $j = i$.
5. If $i < n$ then set $i = i + 1$ and go to step 3.
6. If $j > 0$ then set $x_{jk} = x_{jk-1} + 1$, else stop (no feasible solution for F exists).
7. If $k < D$ then set $k = k + 1$ and go to step 2, else stop.

Steiner and Yeomans prove in [221] that a feasible schedule exists only if $F \geq 1 - \rho_{max}$, where $\rho_{max} = \max_i\{\rho_i\}$ is the biggest production rate of all products. Moreover, they show that a feasible schedule for $F = 1$ always exists. Brauner and Crama prove in [34] the following theorem improving the upper bound given by Steiner and Yeomans.

Theorem 5.13 ([34]). *There exists a feasible schedule for at least one of the values F in the following set*

$$\left\{ \frac{D - \delta_{max}}{D}, \frac{D - \delta_{max} + 1}{D}, \ldots, \frac{D - 1}{D} \right\}. \tag{5.46}$$

Concluding, we have

$$\min \max_{ik} |x_{ik} - k\rho_i| \leq 1 - \frac{1}{D}. \tag{5.47}$$

Recall that a better upper bound is given by Tijdeman (see equation (5.38)). Of course, if there is a feasible schedule for F, then there is a feasible schedule for any $F' \geq F$. Thus, in order to find the smallest possible value of F, Algorithm 5.12 is repeated at most for all values F in set (5.46). The number of possible values of F does not exceed $\delta_{max} = \max_i\{\delta_i\}$. If binary search is performed, finding an optimal solution of the optimization problem requires solving $O(\log \delta_{max})$ decision problems, so the complexity of the algorithm is $O(D \log D)$.

We illustrate the Steiner-Yeomans algorithm by solving the problem given in Example 5.10. Let us check if there exists a schedule for $F = 0.6$. According to Theorem 5.13, 0.6 is the smallest possible value of F. Values $E(i,j)$ and $L(i,j)$ are given in Table 5.4.

Table 5.4. Values $E(i,j)$ and $L(i,j)$ for Example 5.10

j	Product 1		Product 2		Product 3		Product 4	
	E	L	E	L	E	L	E	L
1	4	16	1	4	2	9	2	6
2	–	–	4	6	7	14	5	9
3	–	–	6	9	–	–	8	13
4	–	–	9	11	–	–	–	–

In Table 5.5 for each product we give the numbers of the units eligible for scheduling in stage k. From the set of eligible pairs (i,j) we choose the one with the minimum value of $L(i,j)$. Since ties occur, more than one sequence may be obtained using this method. Our problem has the following solutions: 2432421342, 2432421324, 2432423142, and 2432423124. Notice that the first sequence is identical to the sequence obtained using the Tijdeman algorithm.

Since there exist feasible schedules for $F = 0.6$, all these schedules constitute optimal solutions of the problem from Example 5.10.

Another property of the PRV problem is formulated by Brauner and Crama in [34] as the *small deviations conjecture*, examined further in [35], and proved by Kubiak in [159]. Let us call an instance of the PRV problem *standard* if $0 < \delta_1 \le \delta_2 \le \ldots \le \delta_n$, $n \ge 2$, and the greatest common divisor (gcd) of $\delta_1, \delta_2, \ldots, \delta_n$, and D is 1, i.e., $gcd(\delta_1, \delta_2, \ldots, \delta_n, D) = 1$. The following theorem holds.

Theorem 5.14 ([159]). *Let $(\delta_1, \ldots, \delta_n)$, with $n > 2$, be a standard instance of the PRV problem (5.36), and let B^* be the optimal value of the objective function for this instance. $B^* \le \frac{1}{2}$ if and only if $\delta_i = 2^{i-1}$, for $i = 1, \ldots, n$, and $B^* = (2^{n-1} - 1)/(2^n - 1)$.*

Moreover, Kubiak showed in [159] that for two products a schedule with $B^* < 1/2$ exists if and only if one of the demands is an even

Table 5.5. Eligible pairs for Example 5.10

k	i				Scheduled
	1	2	3	4	product(s)
1	–	1	–	–	2
2	–	1	1	1	4
3	–	1	1	1	3
4	1	1,2	1	1	2
5	1	2	1	1,2	4
6	1	2,3	1	1,2	2
7	1	3	1,2	2	1,3
8	1	3	1,2	2,3	3,1
9	1	3,4	1,2	2,3	2,4
10	1	4	2	3	4,2

number and the other one is an odd number. Thus, for $n = 2$ we have an infinite number of instances with $B^* < 1/2$, while for any $n \geq 3$ there is only one such instance.

5.2.3 Minimizing the total deviation

The most common cost functions f_i in objective (5.32) are the absolute value, which leads to the objective expressed by formula (5.48), and the quadratic function for which the objective is defined by formula (5.49):

$$\text{minimize} \qquad \sum_{i=1}^{n}\sum_{k=1}^{D} |(x_{ik} - k\rho_i| \tag{5.48}$$

$$\text{minimize} \qquad \sum_{i=1}^{n}\sum_{k=1}^{D} (x_{ik} - k\rho_i)^2. \tag{5.49}$$

Miltenburg in [186] observed that minimizing objectives (5.48) and (5.49) results in similar schedules. This observation was first confirmed by computational experiments reported by Kovalyov and Kubiak in [152], and later proved by Corominas and Moreno in [75]. Before formulating the theorems, let us define the *oneness property*. A schedule S has the oneness property if and only if $-1 \leq (x_{ik} - k\rho_i) \leq 1$ for all $i = 1, \ldots, n$ and $k = 1, \ldots, D$. We denote by $\mathcal{X}$ the set of all feasible solutions with the oneness property.

Corominas and Moreno prove the following theorems.

Theorem 5.15 ([75]). *The PRV problems with objectives defined by (5.48) and (5.49) have the same set of optimal solutions on* $\mathcal{X}$.

Theorem 5.16 ([75]). *If an optimal schedule for the RRV problem with objective (5.48) possesses the oneness property, it is also optimal for the PRV problem with objective (5.49), and all optimal solutions of the later one belong to $\mathcal{X}$.*

Concluding, if an instance of the PRV problem with objective (5.49) has no optimal solutions in $\mathcal{X}$, then the PRV problem with objective (5.48) does not have any optimal solutions in $\mathcal{X}$, either. So, in many cases a solution to the problem with objective (5.48) is also a solution to the problem with objective (5.49).

The conjecture, formulated in [152], that convex PRV problems with the objective defined by formula (5.32) are equivalent for any symmetric functions $f = f_i$, $i = 1, \dots, n$, is not true. The proof was given by Corominas and Moreno in [75].

There exist several approaches to solving the PRV problem with the objective to minimize the total deviation. Miltenburg in [186] develops an enumeration algorithm of exponential complexity as well as two heuristics for objective (5.49). Inman and Bulfin in [125] propose a much faster heuristic. However, for some instances the Miltenburg heuristic finds better solutions than theirs. Heuristic algorithms to solve the min-sum PRV problem are also proposed, in [90, 91, 106, 187, 189, 188, 226], and [225]. A dynamic programming algorithm which is exponential with respect to the number of products is presented by Miltenburg et al. in [189]. Finally, Kubiak and Sethi in [162], and [163] propose transformation of the min-sum PRV problem to the assignment problem which is known to be solvable in polynomial time. We present the Inman-Bulfin heuristic and the Kubiak-Sethi algorithm later in this section.

In the previous sections we defined an ideal cumulative production volume of each product i, $i = 1, \dots, n$, in each stage k, $k = 1, \dots, D$. The objective is then to minimize the deviation of the cumulative production volume of product i in stage k from the ideal volume. However, the problem may be also formulated using the concept of the ideal completion times. It may be observed (see Figure 5.1) that the ideal completion time of unit j, $j = 1, \dots, \delta_i$, of product i, $i = 1, \dots, n$ is the intersection point of the graphs obtained for units $j - 1$ and j, i.e.

$$\tau_{ij} = \frac{2j - 1}{2\rho_i} \tag{5.50}$$

Thus the objective is to minimize the deviation of the completion time C_{ij} of unit j of product i in schedule S from its ideal completion time. The following objective functions are considered:

$$\sum_{i=1}^{n}\sum_{j=1}^{\delta_i}(C_{ij}-\tau_{ij})^2 \tag{5.51}$$

and

$$\sum_{i=1}^{n}\sum_{j=1}^{\delta_i}|C_{ij}-\tau_{ij}|, \tag{5.52}$$

where τ_{ij}, defined by (5.50), is the ideal position of unit j of product i, $i = 1, \ldots, n$, $j = 1, \ldots, \delta_i$ in the sequence.

Inman and Bulfin in [125] present an optimization algorithm to solve the problem with the objective functions (5.51) and (5.52). Although their algorithm is not optimal for the min-sum PRV problem, it can be used as an efficient heuristic for that problem, since its computational complexity is $O(nD)$. The algorithm proposed by Inman and Bulfin simply sorts the units of particular products according to the non-decreasing order of values τ_{ij}, and schedules them in that order.

Algorithm 5.17 (Inman-Bulfin [125]).

1. Let $\mathcal{J}$ be the set of all pairs (i, j), such that $i = 1, \ldots, n$, $j = 1, \ldots, \delta_i$.
2. For all (i, j) from $\mathcal{J}$ calculate τ_{ij} according to formula (5.50).
3. Set $k = 1$.
4. Select from $\mathcal{J}$ the pair (i, j) with the minimum value of τ_{ij}. Schedule unit j of product i in stage k. Remove the pair (i, j) from the set $\mathcal{J}$.
5. If $k < D$ then set $k = k + 1$ and go to step 4, else stop.

For illustration, let us use the Inman-Bulfin algorithm to find a schedule of the instance defined in Example 5.10. In Table 5.6 we present the ideal completion times τ_{ij} for all pairs (i, j), $i = 1, 2, 3, 4$, and $j = \delta_i$. There are two optimal solutions (a tie occurs in stage 5): 2432142342 and 2432412342.

Another algorithm for solving the min-sum PRV problem is proposed by Kubiak and Sethi in [162] and [163]. The algorithm is based on the following transformation of the min-sum PRV problem to the assignment problem. The transformation may be applied for any convex, symmetric, nonnegative functions f_i, $i = 1, \ldots, n$, with $f_i(0) = 0$.

Let c^i_{jk} be the cost of scheduling unit j of product i in stage k. The D x D matrix $[c^i_{jk}]$ is the cost matrix for the assignment problem. Below we present how to calculate the values c^i_{jk}.

Table 5.6. Ideal completion times τ_{ij}

j	i			
	1	2	3	4
1	5	1.25	2.5	1.67
2	–	3.75	7.5	5
3	–	6.25	–	8.33
4	–	8.75	–	–

Let us denote by $\varphi_i(j,k)$ the value of function f_i, if unit j of product i is scheduled in stage k, i.e. $\varphi_i(j,k) = f_i(j - k\rho_i)$. Thus, the cost of scheduling product i equals

$$
\begin{aligned}
&\sum_{k=1}^{D} f_i(x_{ik} - k\rho_i) \\
&= \sum_{k=0}^{C_{i1}-1} \varphi_i(0,k) + \sum_{k=C_{i1}}^{C_{i2}-1} \varphi_i(1,k) + \ldots + \sum_{k=C_{i\delta_i}}^{D} \varphi_i(\delta_i,k) \\
&= \sum_{j=0}^{\delta_i} \sum_{k=C_{ij}}^{C_{ij-1}-1} \varphi_i(j,k)
\end{aligned} \tag{5.53}
$$

where $C_{i0} = 0$, $C_{i\delta_i+1} - 1 = D$, and $\varphi_i(0,0) = \varphi_i(\delta_i, D) = 0$. The nonnegative integer variables C_{ij} denote the completion time of unit j of product i and should satisfy the following constraints

$$C_{ij+1} \geq C_{ij} + 1, \quad j = 1, \ldots, \delta_i \tag{5.54}$$

and

$$1 \leq C_{ij} \leq D, \quad j = 1, \ldots, \delta_i. \tag{5.55}$$

Kubiak and Sethi prove in [163] that function (5.53) attains its minimum, under constraints (5.54) and (5.55), for C^*_{ij} defined as

$$C^*_{ij} = \left\lceil \frac{2j-1}{2\rho_i} \right\rceil, \qquad i = 1, \ldots, n. \tag{5.56}$$

We can express the cost of scheduling unit j of product i in stage k as the additional cost incurred when j units of product i are produced by stage k instead of $j-1$ units. If unit j of product i is completed in stage $l < C^*_{ij}$, then $\varphi_i(j,l) \geq \varphi_i(j-1,l)$, otherwise $\varphi_i(j-1,l) \geq \varphi_i(j,l)$. We define

$$\Psi^i_{jl} = \begin{cases} \varphi_i(j - l\rho_i) - \varphi_i(j - 1 - l\rho_i) & \text{if } l < C^*_{ij} \\ \varphi_i(j - 1 - l\rho_i) - \varphi_i(j - l\rho_i) & \text{if } l \geq C^*_{ij} \end{cases} \tag{5.57}$$

Values Ψ^i_{jl} for product $i = 2$ from Example 5.10 and $j = 1$ are presented in Figure 5.2.

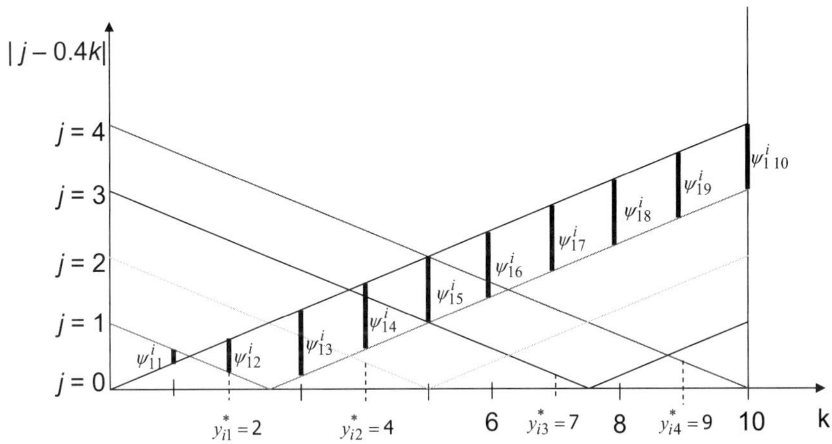

Fig. 5.2. The additional cost Ψ^i_{jl} for $j = 1$.

In order to calculate the cost $c^i_{jk} \geq 0$ of scheduling the j-th unit of product i in stage k it is necessary to find sums of ψ^i_{jk} in the relevant intervals:

$$c^i_{jk} = \begin{cases} \sum_{l=k}^{C^*_{ij}-1} \psi^i_{jl} & \text{if } k < C^*_{ij} \\ 0 & \text{if } k = C^*_{ij} \\ \sum_{l=C^*_{ij}}^{k-1} \psi^i_{jl} & \text{if } k > C^*_{ij} \end{cases} \tag{5.58}$$

A feasible solution to the PRV problem can be defined as a D-element set $Y \subseteq \{(i,j,k) : i = 1,\ldots,n; j = 1,\ldots,\delta_i; k = 1,\ldots,D\}$ such that

(a) For each $k = 1,\ldots,D$, there is exactly one pair $(i,j), i = 1,\ldots,n; j = 1,\ldots,\delta_i$ such that $(i,j,k) \in Y$, i.e. exactly one unit of a product is assigned to each stage.

(b) For each $(i,j), i = 1,\ldots,n; j = 1,\ldots,\delta_i$, there is exactly one $k, k = 1,\ldots,D$, such that $(i,j,k) \in Y$, i.e. each unit j of product i is assigned to exactly one stage.
(c) If $(i,j,k),(i',j',k') \in Y$ and $k < k'$ then $j < j'$, i.e. units with lower numbers are produced earlier.

Let us consider a set Y consisting of D triples (i,j,k), satisfying constraints (a), (b) and (c). A feasible sequence is obtained by scheduling the j-th unit of product i in stage k iff $(i,j,k) \in Y$. The following theorems hold.

Theorem 5.18 ([163]). *If Y is feasible, then*

$$\sum_{i=1}^{n}\sum_{k=1}^{D} f_i(x_{ik} - k\rho_i) = \sum_{(i,j,k\in Y)} c^i_{jk} + \sum_{i=1}^{n}\sum_{k=1}^{D} \inf_j f_i(j - k\rho_i).$$

Theorem 5.19 ([163]). *If Y satisfies constraints (a) and (b), then Y^* satisfying (a), (b) and (c), such that $\sum_{(i,j,k)\in Y} c^i_{jk} \geq \sum_{(i,j,k)\in Y^*} c^i_{jk}$, can be determined in $O(D)$ steps. Moreover, product i is assigned to the same stages in schedules defined by Y and Y^*.*

Since the term $\sum_{i=1}^{n}\sum_{k=1}^{D} \inf_j f_i(j - k\rho_i)$ is independent of the set Y, an optimal set satisfying constraints (a) and (b) can be found by solving the following assignment problem:

$$\text{minimize} \sum_{i=1}^{n}\sum_{j=1}^{\delta_l}\sum_{k=1}^{D} c^i_{jk} x^i_{jk} \tag{5.59}$$

subject to the constraints

$$\sum_{k=1}^{D} x^i_{jk} = 1, \text{ for } i = 1,\ldots,n; j = 1,\ldots,\delta_i, \tag{5.60}$$

and

$$\sum_{i=1}^{n}\sum_{j=1}^{\delta_i} x^i_{jk} = 1, \text{ for } k = 1,\ldots,D, \tag{5.61}$$

where

$$x^i_{jk} = \begin{cases} 1 & \text{if } (i,j) \text{ is assigned to stage } k, \\ 0 & \text{otherwise.} \end{cases}$$

It the solution to the assignment problem (5.59)-(5.61) does not satisfy constraint (c), then by Theorem 5.19, a solution satisfying constraint (c) may be found in $O(D)$ time. Thus the PRV problem can be solved in $O(D^3)$ time. Summarizing, the Kubiak-Sethi algorithm may be formulated as follows.

Algorithm 5.20 (Kubiak-Sethi [162, 163]).

1. For $i = 1, \ldots, n$, $j = 1, \ldots, \delta_i$, and $k = 1, \ldots, D$, calculate values c^i_{jk} according to formula (5.58).
2. Find the solution x^{i*}_{jk} to the assignment problem with cost matrix $[c^i_{jk}]$.
3. For $i = 1, \ldots, n$, $k = 1, \ldots, D$, set $X^{i*}_k = \sum_{l=1}^{k} \sum_{j=1}^{\delta_i} x^{i*}_{jl}$.
4. For $i = 1, \ldots, n$, if $X^{i*}_1 = 1$ then set $C_{i1} = 1$.
5. Set $k = 2$.
6. For $i = 1, \ldots, n$, if $X^{i*}_{k-1} + 1 = X^{i*}_k$ then set $j = X^{i*}_k$ and $C_{ij} = k$.
7. If $k < D$ then set $k = k + 1$, and go to step 6 else stop.

Table 5.7. Values ψ^i_{jk} for Example 5.10

		k										
i	j	1	2	3	4	5	6	7	8	9	10	C^*_{ij}
1	1	0.8	0.6	0.4	0.2	0	0.2	0.4	0.6	0.8	1	5
2	1	0.2	0.6	1	1	1	1	1	1	1	1	2
2	2	1	1	0.6	0.2	1	1	1	1	1	1	4
2	3	1	1	1	1	1	0.2	0.6	1	1	1	7
2	4	1	1	1	1	1	1	1	0.6	0.2	1	9
3	1	0.6	0.2	0.2	0.6	1	1	1	1	1	1	3
3	2	1	1	1	1	1	0.6	0.2	0.2	0.6	1	8
4	1	0.4	0.2	0.8	1	1	1	1	1	1	1	2
4	2	1	1	1	0.6	0	0.6	1	1	1	1	5
4	2	1	1	1	1	1	1	0.8	0.2	0.4	1	9

In order to illustrate the algorithm we find a schedule for the PRV problem defined in Example 5.10 and objective function (5.52). The values ψ^i_{jk} and c^i_{jk} are presented in Tables 5.7 and 5.8, respectively. One of the solutions of the assignment problem appears in boldface in Table 5.8. In Table 5.9 we present the values X^{i*}_k calculated for this solution, and the corresponding optimal schedule represented by pairs (i, j), where i is the number of product and j is the number of unit. Since there are two optimal solutions to the assignment problem, two

Table 5.8. Values c^i_{jk} for Example 5.10

i	j	k = 1	2	3	4	5	6	7	8	9	10
1	1	2	1.2	0.6	0.2	**0**	0	0.2	0.6	1.2	2
2	1	**0.2**	0	0.6	1.6	2.6	3.6	4.6	5.6	6.6	7.6
2	2	2.6	1.6	0.6	**0**	0.2	1.2	2.2	3.2	4.2	5.2
2	3	5.2	4.2	3.2	2.2	1.2	0.2	**0**	0.6	1.6	2.6
2	4	7.6	6.6	5.6	4.6	3.6	2.6	1.6	0.6	0	**0.2**
3	1	0.8	0.2	**0**	0.2	0.8	1.8	2.8	3.8	4.8	5.8
3	2	5.8	4.8	3.8	2.8	1.8	0.8	0.2	**0**	0.2	0.8
4	1	0.4	**0**	0.2	1	2	3	4	5	6	7
4	2	3.6	2.6	1.6	0.6	**0**	0	0.6	1.6	2.6	3.6
4	2	7	6	5	4	3	2	1	0.2	**0**	0.4

Table 5.9. Values $X^{i^*}_k$ for the solution presented in Table 5.8

i	k = 1	2	3	4	5	6	7	8	9	10
1	0	0	0	0	**1**	1	1	1	1	1
2	**1**	1	1	**2**	2	2	**3**	3	3	**4**
3	0	0	**1**	1	1	1	1	**2**	2	2
4	0	**1**	1	1	1	**2**	2	2	**3**	3
(i,j)	(2,1)	(4,1)	(3,1)	(2,2)	(1,1)	(4,2)	(2,3)	(3,2)	(4,3)	(2,4)

sequences optimal for the min-sum PRV problem may be constructed: 2432142342 and 2432412342.

Moreno and Corominas in [192] propose an efficient algorithm for solving the assignment problem (5.59) using the properties of the coefficient matrix obtained by the transformation from the PRV problem. Instances with $n = 10$ and $D = 10000$ have been successfully solved using this algorithm.

Steiner and Yeomans in [222] consider a bicriteria problem with the min-sum and min-max criteria. They show an algorithm to find a Pareto optimal solution in $O(nD^2 logD)$ time. The problem is transformed to the problem of finding a perfect matching in a complete bipartite graph with minimum sum of weights.

5.2.4 Cyclic sequences

In this section we discuss a very important property of the PRV problem, which is the existence of cyclic schedules.

Let σ be a sequence of elements (e.g. integers). The sequence σ^m, obtained as concatenation of m sequences σ is called a *cyclic sequence*. It follows from the previous sections that the time complexity of the PRV scheduling problem, given as a function of the number of products n, remains open. The complexity of all the existing algorithms developed for this problem depends on the value of the total demand D, and thus it is pseudopolynomial in n. For this reason, an important question is whether cyclic sequences are optimal. If the answer is affirmative, a very practical conclusion is that in order to solve the PRV problem with D stages it may be sufficient to find an optimal sequence of length (D/m) and repeat it m times. Thus the computational complexity of the PRV algorithms depends on the existence of cyclic sequences.

Miltenburg in [186] and Miltenburg and Sinnamon in [188] observe the existence of cyclic schedules both for single and multi-level problems and both types of optimality criteria, i.e. maximum and total deviation.

Cyclic sequences have the following property, observed by Miltenburg (see [186]). Let us consider two sequences, σ_1 and σ_2, for the instances $\beta\delta_1, \ldots, \beta\delta_n$ and $\gamma\delta_1, \ldots, \gamma\delta_n$, respectively, where β and γ are positive integers. Then the total deviation for $\sigma_1\sigma_2$ being a concatenation of σ_1 and σ_2 is the sum of total deviations for sequences σ_1 and σ_2.

The question whether a cyclic schedule exists for a given PRV problem is formulated as follows. Let σ be an optimal sequence for the PRV problem with the demand vector $[\delta_1, \ldots, \delta_n]$, and let σ^m be a concatenation of m sequences σ. Is the sequence σ^m, for any $m \geq 1$, optimal for the PRV problem with the demand vector $[m\delta_1, \ldots, m\delta_n]$?

Bautista et al. in [24] prove that a cyclic solution of the single-level min-sum problem exists, if $f_i = f$ for $i = 1, \ldots, n$ and f is a convex and symmetric function with $f(0) = 0$. Kubiak and Kovalyov in [161] prove that if $f_i(x) = f(x)$ for $i = 1, \ldots, n$, with $x \in (0, 1)$ for a convex and symmetric function f, then the cyclic schedule for min-sum problem is optimal. They also give an example which shows that if at least one function is asymmetric, the property does not hold.

The following theorem proved by Kubiak in [157] generalizes the earlier results. It shows that optimal cyclic sequences for the single level min-sum scheduling problems exist for any convex, symmetric and nonnegative functions f_i, $i = 1, \ldots, n$.

Theorem 5.21 ([157]). *Let σ be an optimal sequence for the min-sum PRV scheduling problem with convex, symmetric and nonnegative functions f_i and demand vector $(\delta_1, \ldots, \delta_n)$. Then σ^m, $m \geq 1$, is optimal for the demand vector $(m\delta_1, \ldots, m\delta_n)$.*

The proof is by construction. It can be shown that having an optimal sequence for the min-sum PRV problem we can always construct a cyclic schedule with the value of the objective function which is not greater than optimum. It is not necessary to build a cyclic schedule by reconstruction of an optimal schedule with the techniques used in the proof. An optimal cyclic schedule can be obtained directly by the algorithm presented below.

Algorithm 5.22 (Construction of an optimal cyclic schedule, [157]).

1. Calculate the greatest common divisor m of $(\delta_1, \ldots, \delta_n)$.
2. Apply the Kubiak-Sethi algorithm (see Section 5.2.3) to obtain an optimal sequence for $(\delta_1/m, \ldots, \delta_n/m)$.
3. Concatenate the sequence m times to construct an optimal sequence for the original demands $(\delta_1, \ldots, \delta_n)$.

Steiner and Yeomans prove in [223] that the set of optimal sequences includes a cyclic sequence for both weighted and non-weighted min-max PRV problems for $f_i(x) = |x|$.

Unfortunately, none of the presented results answers the question if there exist an algorithm solving the PRV problem with time complexity bounded by a polynomial function of $\log D$ and n. Concluding, the complexity of the PRV problem remains open.

5.2.5 Transformation of the PRV problem to the apportionment problem

According to [22], any PRV problem can be transformed to a corresponding apportionment problem as follows. Given a PRV problem with n products and demands $\delta_1, \ldots, \delta_n$, let us consider the state when the decision to be taken is which product should be scheduled in stage k. The corresponding apportionment problem is to assign $h = k$ seats to n states with populations $\pi_i = \delta_i, i = 1, \ldots, n$. The cumulative production volume x_{ik} of product i in stage k is equal to the number of seats a_i assigned to state i in a parliament of size $h = k$.

Using this transformation, each PRV problem can be solved by applying any apportionment method (see [158, 134]). Conversely, any apportionment problem may be transformed to a PRV problem as follows.

Let $\boldsymbol{\pi}$ be an n-element vector of populations and h be a house size in an apportionment problem. The corresponding PRV problem is to find a schedule of n products with demands $\delta_i = \pi_i$ in stage $k = h$. Thus, any PRV algorithm is also an apportionment method and can be characterized by properties such as staying within the quota, or population monotonicity. Since in any feasible solution of the PRV problem $x_{ijk-1} < x_{ijk}$, $k = 1, \ldots, D$, any algorithm solving the PRV problem is house monotone.

Example 5.23. Let us consider an apportionment problem with $n = 4$ states, $\pi_1 = 10, \pi_2 = 20, \pi_3 = 30, \pi_4 = 40$, and $h = 5$. The corresponding PRV problem is to construct a schedule of $n = 4$ products with demands $\delta_1 = 10, \delta_2 = 20, \delta_3 = 30, \delta_4 = 40$, in stages 1 through 5. We can apply, for example, the Tijdeman algorithm and obtain the sequence 43243. The corresponding solution to the original apportionment problem is $a_1 = 0, a_2 = 1, a_3 = 2, a_4 = 2$.

Since the Tijdeman algorithm guarantees that $|x_{ik} - k\rho_i| < 1$, it is obvious that it stays within the quota. Le us recall that the Tijdeman algorithm schedules in stage k product i', such that

$$\frac{1 - \frac{1}{2n-2} - k\rho_{i'} + x_{i'k-1}}{\rho_{i'}} = \min_i \left\{ \frac{1 - \frac{1}{2n-2} - k\rho_i + x_{ik-1}}{\rho_i} \right\}$$
$$= \max_i \left\{ \frac{\delta_i}{1 - \frac{1}{2n-2} + x_{ik-1}} \right\}.$$

In terms of the corresponding apportionment problem, the last equation means that the k-th seat in the parliament is assigned to state i', such that

$$\frac{\pi_{i'}}{1 - \frac{1}{2n-2} + a_{i'}} = \max_i \left\{ \frac{\pi_i}{1 - \frac{1}{2n-2} + a_i} \right\}.$$

Such assignment is equivalent to the assignment generated by a divisor method with the divisor criterion equal to $d(a) = a + 1 - 1/(2n-2)$. Although the selection criterion is the same as in the divisor method, the eligible set defined by the Tijdeman algorithm is a proper subset of the eligible set defined by the quota divisor method with $d(a) = a + 1 - 1/(2n-2)$. It means that product i may be eligible in stage k according to the quota divisor method, but not eligible according to the Tijdeman algorithm. The proof presented in [134] is based on an example with $\boldsymbol{\delta} = [2, 3, 7]$. In order to stress that the eligible set in the

Tijdeman algorithm is not identical with the eligible set in the quota-divisor method with $d(a) = a+1-1/(2n-2)$, we say that the Tijdeman algorithm is a *quasi* quota-divisor method as stated in Theorem 5.24 (see [134]).

Theorem 5.24 ([134]). *The Tijdeman algorithm is a quasi quota-divisor method.*

It is also an obvious result that the Steiner-Yeomans algorithm is house monotone. As we have noticed, ties may occur in the Steiner-Yeomans algorithm. If the ties are broken arbitrarily, no more properties of the algorithm can be proved. Thus, let us assume that in case of a tie we choose (i, j) with the smallest $(j + F)/\rho_i$. Under this assumption the Steiner-Yeomans algorithm is also a quota-divisor method. The following theorem stating this property was proved in [134].

Theorem 5.25 ([134]). *The Steiner-Yeomans algorithm with $F < 1$ and a tie $L(i,j) = L(i',j')$ between units (i,j) and (i',j') broken by choosing the unit with*

$$\min\left\{\frac{1}{\rho_i}(j+F), \frac{1}{\rho_i'}(j'+F)\right\}$$

is a quota-divisor method with divisor criterion

$$d(a) = a + F.$$

The Steiner-Yeomans algorithm finds schedules with $\max\{|x_{ik} - \rho_i|\} \le F < 1$, so obviously it stays within the quota. Moreover, in each stage k it schedules a unit of product i', such that

$$\frac{x_{i'k-1} + F}{\rho_{i'}} = \min_i \left\{\frac{x_{ik-1} + F}{\rho_i}\right\} = \max_i \left\{\frac{\rho_i}{x_{ik-1} + F}\right\}. \tag{5.62}$$

In terms of the corresponding apportionment problem, the above equation means that the k-th seat in the parliament is assigned to the state with the maximum value of $\pi_i/(a_i + F)$. Such assignment is equivalent to the assignment generated by a divisor method with the divisor criterion equal to $d(a) = a+F$. Concluding, the Steiner-Yeomans algorithm with the above formulated tie-breaking rule is a quota-divisor method.

The Tijdeman algorithm and the Steiner-Yeomans algorithm have been developed to solve the min-max PRV problem. Thus, it is obvious that both stay within the quota. Now we examine the Inman-Bulfin and the Kubiak-Sethi algorithms, both minimizing the min-sum criteria.

It was observed in [22] that the Inman-Bulfin algorithm is equivalent to the Webster divisor method. Consequently, the Inman-Bulfin algorithm as a parametric divisor method is uniform and does not stay within the quota (see Section 2.2). Therefore, the following theorem holds.

Theorem 5.26 ([134]). *The Inman-Bulfin algorithm is uniform and does not stay within the quota.*

Finally, let us consider the Kubiak-Sethi algorithm. The following properties hold.

Theorem 5.27 ([134]). *The Kubiak-Sethi algorithm*

- *is house monotone,*
- *does not stay within the quota,*
- *is not uniform, and*
- *is not population monotone.*

We have noticed earlier in this section that all algorithms solving the PRV problem are house monotone, so the first property obviously holds. The second property follows immediately from an observation by Corominas and Moreno [75] that no solution minimizing the function

$$\sum_{i=1}^{n}\sum_{k=1}^{D}|x_{ik} - k\rho_i|$$

stays within the quota for the instance with $n = 6$ and $\boldsymbol{\delta} = [23, 23, 1, 1, 1, 1]$. An instance with a non-uniform solution obtained using the Kubiak-Sethi algorithm presented in [134] proves the third property. This instance consists of $n = 5$ products with the demand vector $\boldsymbol{\delta} = [7, 6, 4, 2, 1]$. The Kubiak-Sethi algorithm has 18 solutions, and one of them is 12312413215231421321. If the algorithm was uniform, the sequence 2324325234232, obtained by removing product 1 from the original sequence, should be obtained by the Kubiak-Sethi algorithm as a solution to the restricted problem with $\boldsymbol{\delta}' = [6, 4, 2, 1]$. However, it is not so. Finally, since all population monotone methods are uniform (see Section 2.2), the Kubiak-Sethi algorithm is not population monotone either.

In Table 5.10 we summarize the results presented in this section. The algorithms developed to solve the PRV problem are classified according to the properties they possess as apportionment methods. In order to present a complete classification we also include the apportionment methods in the table, and we highlight the PRV algorithms with boldface.

5.3 Scheduling periodic tasks

A brief characteristics of scheduling problems occurring in real-time systems is presented in Chapter 2.1. In this section we formulate the problem of scheduling periodic tasks in a hard real-time environment. This problem was first formulated by Liu and Layland in [177]. It is considered one of the most general formulations of the hard real-time scheduling problem.

Many computer systems operate in the so-called real-time environment. Most of real-time applications can be found in control and monitoring of industrial processes. Usually, a single computer performs several tasks executed in response to events in the system controlled or monitored by the computer. None of the tasks may be executed before the occurrence of the event that requests it. Each task has to be completed within some fixed time after the request occurs. The demand for service in the interval of a given length starting from the request characterizes the hard real-time system [180], in contrast to soft real-time system, where a statistical distribution of response times is acceptable. A general characteristics of the hard real-time scheduling problems of this type was first presented by Liu and Layland [177]. They formulate some properties of schedulling algorithms and propose two solution approaches: one with fixed and one with dynamic task priorities. In this section we show that the Liu-Layland problem can be transformed to the apportionment problem and we provide necessary and sufficient conditions for a divisor method to solve the Liu-Layland problem. In Section 5.3.1 we formulate the problem and define its transformation to the PRV problem as well as to the apportionment problem, then in Section 5.3.2 we present the solution algorithms, and finally, in Section 5.3.3, we discuss the properties of feasible schedules.

5.3.1 Problem formulation

The features characterizing the real-time systems are summarized in [177] by Liu and Layland, who formulate the following assumptions about program behavior in a hard real-time environment:

- the requests for all tasks for which hard deadlines exist are periodic, with constant intervals between requests;

- deadlines result from run-ability only – i.e. each task must be completed before the next request for it occurs;
- the tasks are independent, which means that the requests for a certain task do not depend on initialization, completion or request of any other task;
- processing time of each task is constant. Processing time is understood as the time needed by a processor to execute this task without interruption;
- any non-periodic tasks in the system are special. There are no deadlines assigned to the non-periodic tasks and they displace periodic tasks when they are being processed.

Based on the above assumptions the Liu-Layland periodic scheduling problem is defined as follows. Let us consider a set of n independent, preemptive tasks with periodic requests (*periodic* tasks) to be scheduled on a single processor. A periodic task i is characterized by its *request period* H_i and processing time p_i, $i = 1, \ldots, n$. The *request rate* of a task is defined as the reciprocal of its request period. It is required that task i is executed for p_i time units in every consecutive period of length H_i. It means that the j-th request for task i that occurs at time $(j-1)H_i$, $j = 1, \ldots$, must be completed by the time jH_i. Missing the deadline may prove fatal for the controlled system.

Without loss of generality we may assume that all numbers are positive integers and $p_1 \leq H_i$ for $i = 1, 2, \ldots, n$. The problem is to find a schedule such that task i is executed exactly p_i time units in each interval $[(j-1)H_i, jH_i]$ for $j = 1, \ldots$. Notice that the Liu-Layland problem is a decision problem, i.e. that the objective is to find a feasible sequence.

Let us illustrate the problem with an example.

Example 5.28. Consider 2 tasks with $p_1 = 2$, $H_1 = 4$, $p_2 = 5$ and $H_2 = 10$.

Figure 5.3 illustrates the execution of two tasks in their request periods. The goal is to find a feasible schedule on a single machine, therefore only one task can be executed in any time unit. A feasible schedule of tasks defined in Example 5.28 is presented in Figure 5.4.

Notice that in fact, in order to obtain a feasible schedule it is enough to find a sequence of numbers $1, \ldots, n$, such that each number occurs exactly p_i times in every subsequence defined by positions $(j-1)H_i, \ldots, jH_i$, for $j = 1, \ldots$. It is easy to notice that if we can find a feasible sequence of length $lcm(H_1, \ldots, H_n)$, where lcm is the least common multiplier, then a concatenation of an arbitrary number of such schedules is a solution to the Liu-Layland problem. A schedule obtained

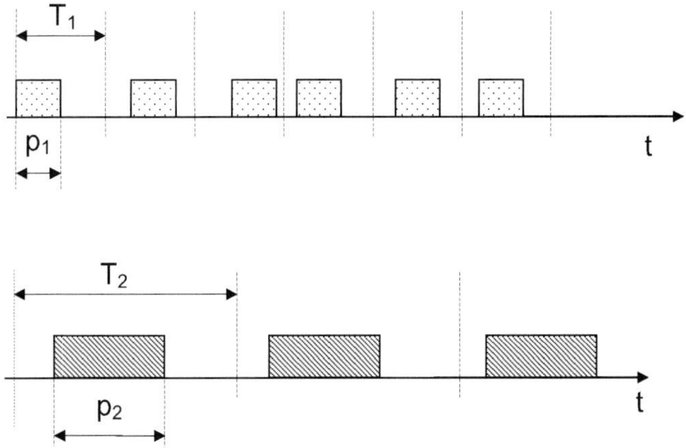

Fig. 5.3. Execution of tasks in their request periods.

this way is a cyclic schedule. The sequence 11221122121211221122 (of length 4*5 = 20 positions) gives a solution to the problem instance defined in Example 5.28.

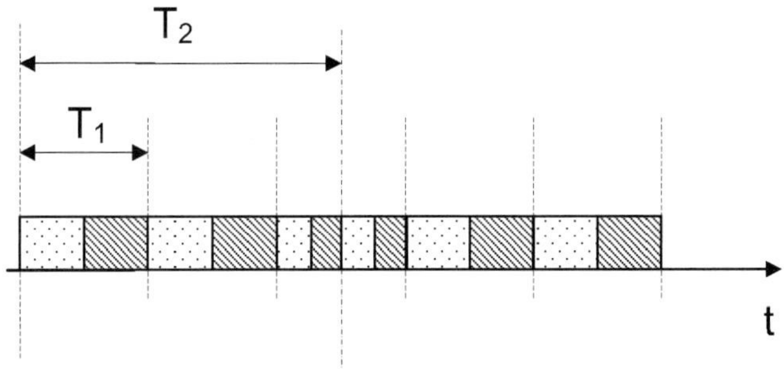

Fig. 5.4. A feasible schedule of tasks from Example 5.28.

5.3.2 Scheduling algorithms

Liu and Layland [177] propose two approaches to solving the periodic scheduling problem. They both are priority driven ones. It means that whenever there is a request for a task that has a higher priority than the one currently executed, the running task is immediately interrupted and the newly requested task is started. In the first approach (*static*) priorities are assigned once to all tasks. In the other one (*dynamic*) the priorities may change from request to request. These two approaches may be combined so that some tasks are given fixed priorities, while the priorities of the remaining ones may change from request to request. Such approach is called *mixed*.

Before we present the algorithms, let us define the processor utilization factor. An important characteristics of a scheduling algorithm solving the Liu-Layland problem is the achievable processor utilization guaranteed by the algorithm. The *processor utilization factor* U is defined as the fraction of processor time spent on the execution of the set of tasks,

$$U = \sum_{i=1}^{n} \frac{p_i}{H_i}.$$

Corresponding to a priority assignment, a set of tasks is said to fully utilize the processor if the priority assignment is feasible for the set and if an increase in the processing time of any task in the set makes the priority assignment infeasible.

Fixed priority scheduling algorithm

Fixed priority assignment means that priorities are assigned to all tasks in the first step of the algorithm and they remain constant in all the following steps.

Let us assume that $0 \leq (j-1)H_i \leq k \leq jH_i \leq lcm(H_1, \ldots, H_n)$ for some integer j. We say that scheduling of task i in time unit k is eligible if the processor time allotted to task i in the interval $[(j-1)H_i, k-1]$ is less than p_i. If, however, the processor time allotted to task i in the interval $[(j-1)H_i + 1, jH_i]$ is less than p_i for any j, then the schedule is not feasible.

Algorithm 5.29 (Fixed priority scheduling algorithm [177]).

1. For each i, $i = 1, \ldots, n$, order tasks according to the priority rule.
2. Set $k = 1$.
3. From the set of tasks eligible for scheduling in time unit k select the one with the highest priority.
4. If $k < lcm(H_i)$, then set $k = k + 1$ and go to step 3.

A priority assignment, where priorities are assigned to tasks according to a non-decreasing request rate independent of their processing times, is called the *rate monotonic priority assignment*. There are many various priority rules that may be applied in the fixed priority scheduling algorithm. Obviously, they result in different schedules. The *rate monotonic priority assignment* is considered optimal among all fixed priority rules. The following theorem explains the meaning of optimality.

Theorem 5.30 ([177]). *If a feasible fixed priority assignment exists for some task set, then the rate monotonic priority assignment is feasible for that task set.*

The fixed priority scheduling algorithm may fail to find a feasible schedule even if such a schedule exists. In order to illustrate such situation let us consider the instance of the Liu-Layland problem defined in Example 5.28 and solve it using the fixed priority scheduling algorithm with the rate monotonic priority assignment. According to this algorithm, task 1 in our example has higher priority than task 2. Thus, in time units 1 and 2 task 1 is scheduled. In time units 3 and 4, the request for task 1 is completed, so the remaining task, task 2, is scheduled. In time unit 5, a new request for task 1 appears, so task 2 is stopped and task one is scheduled in time unit 5 and 6. In time units 7 and 8 task 2 may again be processed. In time unit 9 the next request for task 1 occurs and task 2 has to be stopped. Task 1 is scheduled in time units 9 and 10. At this moment we can see that the schedule is not feasible, since the processor time allotted to task 2 in interval $[0, H_2]$ is $4 < 5 = p_2$. This schedule is presented in Figure 5.5a.

For a given fixed priority scheduling algorithm, the *least upper bound* of the utilization factor is the minimum of the utilization factors over all sets of tasks that fully utilize the processor. For all task sets whose processor utilization factor is below that bound, there exists a fixed priority assignment which is feasible. It is proved in [177] and [86] that the least upper bound depends on the number of tasks, as stated in Theorem 5.31.

Theorem 5.31 ([177]). *For a set of n tasks with fixed priority order, the least upper bound to the processor utilization factor is $U = n(\sqrt[n]{2} - 1)$.*

Thus the least upper bound imposed on the processor utilization factor can approach $ln2$ for a large task set. A better solution can be found applying the dynamic priority assignment, presented in the next section.

Dynamic priority assignment

Liu and Layland show that dynamic priority assignment leads to better processor utilization than the fixed priority assignment. The algorithm proposed by Liu and Layland in [177] based on dynamic priorities is called *deadline driven scheduling algorithm.* Priorities are assigned to tasks according to the deadlines of their current requests, which means that a task is assigned the highest priority if the deadline of its current request is the nearest. At any time, the task with the highest priority and yet unfulfilled request is executed. Thus, contrary to the fixed priority algorithm, the priorities change in time. Let us denote by x_{ik} the number of time units assigned to task i up to time k.

Algorithm 5.32 (Deadline driven scheduling algorithm [177]).

1. Set $k = 1$.
2. Assign time unit k to the task i with the closest due date of the incomplete request, i.e., with $x_{ik} < j_i p_i$, where $(j_i - 1)H_i < j \leq j_i H_i$, for the integer j_i, which denotes the number of request periods of task i completed by time k.
3. If $k < lcm(H_i)$ then set $k = k + 1$ and go to step 2.

The deadline driven algorithm is optimal among the algorithms solving the Liu-Layland scheduling problem. Concluding, the least utilization factor is uniformly 100 percent, and the following necessary and sufficient condition for a set of tasks to have a periodic schedule may be formulated.

Theorem 5.33 ([177]). *For a given set of n tasks, the deadline driven scheduling algorithm is feasible if and only if*

$$\sum_{i=1}^{n} \frac{p_i}{H_i} \leq 1. \tag{5.63}$$

The following example illustrates the differences between schedules obtained according to the algorithms described above.

Let us consider again Example 5.28. Figure 5.5 shows the schedules obtained by the fixed and dynamic priority algorithms. Observe that the schedule obtained by the rate monotonic approach is not feasible since the due date of the second task is violated. The schedule obtained by the deadline driven algorithm is feasible.

As the third approach, Liu and Layland propose the so-called *mixed scheduling algorithm.* The idea of this algorithm is to divide the set of tasks into two groups. One group consists of $n_1 < n$ tasks with the shortest processing times which are scheduled according to the fixed priority assignment. The remaining tasks are scheduled according to the deadline driven scheduling algorithm when the processor is not occupied by tasks from the first group.

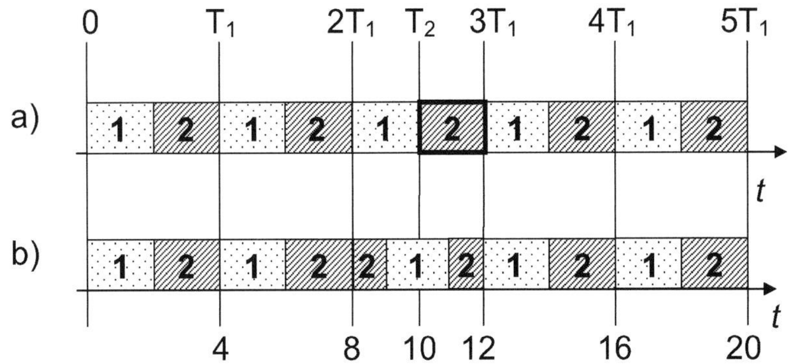

Fig. 5.5. Schedules obtained by fixed and dynamic priority driven algoritms: a) rate-monotonic algorithm, b) deadline driven algorithm.

5.3.3 Properties of feasible schedules

In this section we discuss some properties of feasible schedules for the Liu-Layland problem on the basis of the apportionment theory. We start with constructing the transformation of the Liu-Layland problem to the apportionment problem. In the previous section we show a transformation of the Product Rate Variation problem to the apportionment problem. Below, we define the transformation of the Liu-Layland problem to the PRV problem. Combining the two, we may transform any instance of the Liu-Layland problem to an instance of the apportionment problem.

A natural transformation of the Liu-Layland problem to the PRV problem may be obtained as follows. Each task in the Liu-Layland problem corresponds to a task in the PRV problem. The schedule cycle time Λ is interpreted as the total demand of all tasks $\sum_{i=1}^{n} \delta_i$, and the total processing time of task i in cycle time Λ, calculated as $(\Lambda p_i)/H_i$, corresponds to the demand δ_i of task i. Notice that

$$\frac{p_i}{H_i} = \frac{\delta_i}{\sum_{i=1}^{n} \delta_i}.$$

Combining the above transformation with the transformation of the PRV problem to the apportionment problem described in Section 5.2.5, we can obtain an instance of the apportionment problem corresponding to any instance of the Liu-Layland problem.

Using the transformation of the Liu-Layland problem to the PRV problem, Kubiak [156] proved that any solution to the PRV problem satisfying quota is a solution to the corresponding Liu-Layland problem. In other words, staying within the quota is a sufficient condition for any house monotone method to solve the Liu-Layland problem. Consequently Kubiak proposed to use the Steiner-Yeomans [221] and the Tijdeman [238] algorithms as well as the quota methods by Balinsky and Young [20] and Still [224] to solve the Liu-Layland problem.

Satisfying quota, however, is not a necessary condition for a house monotone method to solve the Liu-Layland problem, as proved in [135]. Namely, consider the following instance of the Liu-Layland problem: $p_1 = 3, H_1 = 5, p_2 = 2, H_2 = 5$, and a house monotone method that schedules tasks according to non-decreasing request period and breaking ties by choosing the task with the longest processing time. The following schedule results: $((1,1),(1,2),(1,3),(2,1),(2,2))$, where (i,j) denotes the jth unit of task i. It is easy to notice that at position 3 the lower quota of task 2 is violated, since

$$\lfloor q_2 \rfloor = \left\lfloor \frac{2*3}{5} \right\rfloor = 1,$$

while $x_{23} = 0$.

It is worth noticing that no divisor method solves the Liu-Layland problem. It follows from the following theorem proved in [135].

Theorem 5.34 ([135]). *Staying within the quota is a necessary condition for any divisor method to solve the Liu-Layland problem.*

Since no divisor method satisfies the quota, an immediate conclusion is that no divisor method solves the Liu-Layland problem.

Let us recall the following propositions proved by Balinski and Young.

Proposition 5.35 ([20]). *The Jefferson method is the only divisor method that stays above the lower quota.*

Proposition 5.36 ([20]). *The Adams method is the only divisor method that stays below the upper quota.*

We illustrate Theorem 5.34 showing that neither the Jefferson nor Adams methods solve the Liu-Layland problem.

Example 5.37. Consider an instance of 4 tasks.

$$H_1 = 5, p_1 = 3, H_2 = 5, p_2 = 1, H_3 = 10, p_3 = 1, H_4 = 10, p_4 = 1.$$

The deadline driven scheduling algorithm proposed by Liu and Layland in [177] gives the following sequence of tasks:

$$1, 1, 1, 2, 3, 1, 1, 1, 2, 4. \tag{5.64}$$

The transformation of the instance to the apportionment problem gives a four state instance with populations

$$\pi_1 = \frac{\Lambda p_1}{H_1} = 6, \pi_2 = \frac{\Lambda p_2}{H_2} = 2, \pi_3 = \frac{\Lambda p_3}{H_3} = 1, \pi_4 = \frac{\Lambda p_4}{H_4} = 1, \tag{5.65}$$

where $\Lambda = lcm(5, 10) = 10$.

The Jefferson method results in the following sequence.

$$1, 1, 1 <-> 2, 1, 1, 1 <-> 2 <-> 3 <-> 4, \tag{5.66}$$

where $x <-> y$ means that the x and y can be interchanged in the sequence. In (5.66) only two (instead of required three) positions between positions 6 and 10 are occupied by task 1. Thus, (5.66) is not a feasible solution of the Liu-Layland problem and we may conclude that the Jefferson method of apportionment does not solve the Liu-Layland problem.

The Adams method results in the following sequence.

$$1, 2, 3 <-> 4, 1, 1, 2 <-> 1, 1, 1 \tag{5.67}$$

for the apportionment problem (5.65). In (5.67) only two (instead of required three) positions between positions 1 and 5 are occupied by task 1. We see that (5.67) is not a feasible solution of the Liu-Layland problem. Thus, the Adams method of apportionment does not solve the Liu-Layland problem, either.

Since no divisor method satisfies the quota, an immediate conclusion is that none of these methods can be applied to scheduling periodic hard real-time tasks. On the other hand, all quota-divisor methods stay within the quota, so they can be applied to find solutions of the Liu-Layland problem. Concluding, no method solving the Liu-Layland problem is population monotone.

The results presented in this chapter show that the apportionment methods provide novel solutions to numerous scheduling problems and offer tools for a deeper analysis of the problem properties.

Table 5.10. Classification of the PRV algorithms

	House monotone methods		Methods that are not house monotone
	Population monotone methods	Methods that are not population monotone	
Methods staying witin the quota	Empty set (see the Impossibility Theorem)	Still method **Tijdeman** **Steiner-Yeomans**	Hamilton method
Methods not staying witin the quota	Uniform = Rank-index methods		
	Divisor methods (Jefferson, Adams, Webster, Dean, Hill method) **Inman-Bulfin**	**Kubiak-Sethi**	

References

1. Abdul-Razaq T, Potts C (1988) Dynamic programming state-space relaxation for single machine scheduling. Journal of the Operational Research Society 39:141–152
2. Achuthan N. R, Grabowski J, Sidney JB (1981) Optimal flow shop scheduling with earliness and tardiness penalties. Opsearch 18:117-138
3. Adamopoulos GI, Pappis CP (1996) Scheduling jobs with different, job-dependent earliness and tardiness penalties using the SKL method. European Journal of Operational Research 88:336–344
4. Adamopoulos GI, Pappis CP (1996) Single machine scheduling with flow allowances. Journal of the Operational Research Society 47:1280–1285
5. Adamopoulos GI, Pappis CP (1998) Scheduling under a common due-date on parallel unrelated machines. European Journal of Operational Research 105:494–501
6. Al-Turki UM, Mittenthal J, Raghavachari M (1996) A dominant subset of V-shaped sequences for a class of single machine sequencing problems. European Journal of Operational Research 88:345–347
7. Alidaee B, Ahmadian A (1993) Two parallel machine sequencing problems involving controllable job processing times. European Journal of Operational Research 70:335–341
8. Alidaee B, Dragan I (1997) A note on minimizing the weighted sum of tardy and early completion penalties in a single machine: A case of small common due date. European Journal of Operational Research 96:559–563
9. Arakawa M, Fuyuki M, Inoue I (2002) A simulation-based production scheduling method for minimizing the due-date deviation. International Transactions in Operational Research 9:153–167
10. Arnold JRT (1998) Introduction to materials management. Prentice Hall, Upper Saddle River NJ
11. Bachman A, Cheng TCE, Janiak A, Ng CT (2002) Scheduling start time dependent jobs to minimize the total weighted completion time. Journal of the Operational Research Society 53:688–693

12. Bagchi U (1989) Simultaneous minimization of mean and variation of flow time and waiting time in single machine systems. Operations Research 37:118–125
13. Bagchi U, Sullivan RS, Chang V-L (1986) Minimizing mean absolute deviation of completion times about a common due date. Naval Research Logistics Quarterly 33:227–240
14. Bagchi U, Sullivan RS, Chang V-L (1987) Minimizing mean squared deviation of completion times about a common due date. Management Science 33:894–906
15. Bagchi U, Chang V-L, Sullivan RS (1987) Minimizing absolute and squared deviations of completion times with different earliness and tardiness penalties and a common due date. Naval Research Logistics 34:739–751
16. Baker KR, Scudder GD (1989) On the assignment of optimal due dates. Journal of the Operational Research Society 40:93–95
17. Baker KR, Scudder GD (1990) Sequencing with earliness and tardiness penalties: a review. Operations Research 38:22–36
18. Balakrishnan N, Kanet JJ, Sridharan SV (1999) Early/tardy scheduling with sequence dependent setups on uniform parallel machines. Computers & Operations Research 26:127–141
19. Balinski ML, Young HF (1975) The quota method of apportionment. American Mathematical Monthly 82:701–730
20. Balinski M, Young HP (1982) Fair Representation: Meeting Ideal One Man, One Vote. Yale University Press, New Haven CT
21. Bauman J, Józefowska J (2006) Minimizing the earliness-tardiness costs on a single machine. Computers & Operations Research 33:3219–3230
22. Bautista J, Companys R, Corominas A (1996) A note on the relation between the Product Rate Variation (PRV) problem and the apportionment problem. Journal of the Operational Research Society 47:1410–1414
23. Bautista J, Companys R, Corominas A (1996) Heuristic and exact algorithms for solving the Monden problem. European Journal of Operational Research, 88: 101-113
24. Bautista J, Companys R, Corominas A (1997) Modelling and solving the Production Rate Variation problem (PRVP). Top. Revista de la Sociedad de Estadística e Investigación Operativa 5:221–239
25. Bector CR, Gupta Y, Gupta MC (1988) Determination of an optimal common due date and optimal sequence in a single machine job shop. International Journal of Production Research 26:613–628
26. Bector CR, Gupta Y, Gupta MC (1989) V-shape property of optimal sequence of jobs about a common due date on a single machine. Computers & Operations Research 16:583–588
27. Biskup D (1999) Single machine scheduling with learning considerations. European Journal of Operational Research 115:173–178

28. Biskup D, Cheng TCE (1999) Single machine scheduling with controllable processing times and earliness, tardiness and completion time penalties. Engineering Optimization 31:329–336
29. Biskup D, Feldmann M (2001) Benchmarks for scheduling on a single machine against restrictive and unrestrictive common due dates. Computers & Operations Research 28:787–801
30. Biskup D, Jahnke H (2001) Common due date assignment for scheduling on a single machine with jointly reducible processing times. International Journal of Production Economics 69:317–322
31. Bitran G, Chang L (1987) A mathematical programming approach to a deterministic Kanban system. Management Science 33:427–441
32. Blum N, Floyd RW, Pratt V, Rivest RL, Tarjan RE (1973) Time bounds for selection. Journal of Computer and System Sciences 7:448–461
33. Błażewicz J, Ecker KH, Pesch E, Schmidt G, Węglarz J (2001) Scheduling Computer and Manufacturing Processes. Springer Verlag, Berlin
34. Brauner N, Crama Y (2004) The maximum deviation just-in-time scheduling problem. Discrete Applied Mathematics 134:25–50
35. Brauner N, Jost V, Kubiak W (2002) On symmetric Fraenkel's and small deviations conjecture. Report #54, Laboratoire Leibniz-IMAG, Grenoble
36. Brucker P (2001) Scheduling Algorithms. Springer Verlag, Berlin
37. Bülbül K, Kaminsky P, Yano C (2004) Preemption in single machine earliness/tardiness scheduling. Submitted for publication
38. Buttazzo GC (1997) Hard Real-Time Computing Systems: Predictable Scheduling Algorithms and Applications. Kluwer, Boston.
39. Cai X (1991) Variance minimization in single machine systems: complexity and algorithm. Research Report, Department of Mathematics, University Western Australia, Perth
40. Cai X (1995) Minimization of agreeably weighted variance in single machine systems. European Journal of Operational Research 85:576–592
41. Cai X (1996) V-shape property for job sequences that minimize the expected completion time variance. European Journal of Operational Research 91:118–123
42. Cai X, Lum VYS, Chan JMT (1997) Scheduling about a common due date with job-dependent assymetric earliness and tardiness penalties. European Journal of Operational Research 98:154–168
43. Cai X, Cheng TCE (1998) Multi-machine scheduling with variance minimization. Discrete Applied Mathematics 84:55–70
44. Čepek O, Sung SC (2005) A quadratic time algorithm to maximize the number of just-in-time jobs on identical parallel machines. Computers & Operations Research 32:3265–3271
45. Chand S, Schneeberger H (1988) Single machine scheduling to minimize weighted earliness subject to no tardy jobs. European Journal of Operational Research 34:221–230

46. Chang PC (1999) A branch and bound approach for single machine scheduling with earliness and tardiness penalties. Computers and Mathematics with Applications 27:133–144
47. Chen ZL (1996) Scheduling and common due date assignment with earliness-tardiness penalties and batch delivery cost. European Journal of Operational Research 93:49–60
48. Chen ZL (1997) Scheduling with batch setup times and earliness-tardiness penalties. European Journal of Operational Research 96:518–537
49. Chen ZL, Lu Q, Tang G (1997) Single machine scheduling with discretely controllable processing times. Operations Research Letters 21:69–76
50. Cheng TCE (1984) Optimal due date determination and sequencing of n jobs on a single machine. Journal of the Operational Research Society 35:433–437
51. Cheng TCE (1985) A duality approach to optimal due date determination. Engineering Optimization 9:127–130
52. Cheng TCE (1987) An algorithm for the CON due date determination and sequencing problem. Computers & Operations Research 14:537–542
53. Cheng TCE (1987) Optimal total-work-content-power due-date determination and sequencing. Computers & Operations Research 14:537–542
54. Cheng TCE (1988) Optimal common due date with limited completion time deviation. Computers and mathematics with Applications 14:579–582
55. Cheng TCE (1989) A heuristic for common due date assignment and job scheduling on parallel machines. Journal of the Operational Research Society 40:1129–1135
56. Cheng TCE (1989) An alternative proof of optimality for the common due-date assignment problem. European Journal of Operational Research 37:250-253 Corrigendum: European Journal of Operational Research 38:259
57. Cheng TCE (1990) A note on partial search algorithm for the single-machine optimal common due date assignment and sequencing problem. Computers & Operations Research 17:321–324
58. Cheng TCE (1990) Common due-date assignment and scheduling for a single processor to minimize the number of tardy jobs. Engineering Optimization 16:129–136
59. Cheng TCE, Chen ZL (1994) Parallel-machine scheduling problems with earliness and tardiness penalties. Journal of the Operational Research Society 45:685–695
60. Cheng TCE, Gupta M (1989) Survey of scheduling research involving due date determination decisions. European Journal of Operational Research 38:156–166
61. Cheng TCE, Janiak A (1992) Scheduling and resource allocation problems in some manufacturing systems. Proc. 8th International Conference on CAD/CAM, Robotics, pp 1657–1671

62. Cheng TCE, Janiak A (1993) Scheduling problem in some production processes. In: Cotsaftis M, Vernadat F (eds) Advances in Factories of the Future. Elsevier, Amsterdam New York
63. Cheng TCE, Janiak A (1994) Resource optimal control in some single-machine scheduling problems. IEEE Transactions on Automatic Control 39:1243–1246
64. Cheng TCE, Kahlbacher HG (1991) A proof of the longest-job-first policy in one-machine scheduling. Naval Research Logistics 38:715–720
65. Cheng TCE, Kang LY, Ng CT (2005) Single machine due-date scheduling of jobs with decreasing start-time dependent processing tomes. International Transactions in Operational Research 12:355–366
66. Cheng TCE, Chen ZL, Li CL (1996) Parallel machine scheduling with controllable processing times. IIE Transactions 28:177–180
67. Cheng TCE, Oguz C, Qi XD (1996) Due-date assignment and single machine scheduling with compressible processing times. International Journal of Production Economics 43:29–35
68. Chu C, Gordon V, (2000) TWK due date determination and scheduling model: NP-hardness and polyomially solvable case. In: Proth J-M, Tanaev V (eds), Proceedings of the International Workshop on Discrete optimization methods in scheduling and computer-aided design, Academy of Sciences of Belarus, Minsk
69. Chretienne P (2001) Minimizing the earliness and tardiness cost of a sequence of tasks on a single machine. Recherche Opérationnelle - RAIRO 35:165–187
70. Coffman EG (1976) Computer and Job Shop Scheduling Theory. John Wiley & Sons, New York
71. Coleman BJ (1992) A simple model for optimizing the single machine early/tardy problem with sequence dependent setups. Production and Operations Management 1:225–228
72. Conway RW (1965) Priority dispatching and job lateness in a job shop. Journal of Industrial Engineering 16:228–237
73. Conway RW, Maxwell WL, Miller LW (1967) Theory of scheduling. Addison-Wesley Publishing Company, Reading MA
74. Cook SA (1971) The complexity of theorem-proving procedures. Proceedings of the 3rd Annual ACM Symposium on Theory of Computing, pp 151–158
75. Corominas A, Moreno N (2003) About the relations between optimal solutions for different types of min-sum balanced JIT optimisation problems. INFOR 41:333–339
76. Cox JF, Blackstone JH, Spencer MS (eds) (1995) APICS Dictionary. APICS. Falls Church VA
77. Davis JS, Kanet JJ (1988) Single machine scheduling with a non-regular convex performance measure. Working Paper, Department of management, Clemson University, Clemson SC

78. Davis JS, Kanet JJ (1993) Single machine scheduling with early and tardy completion costs. Naval Research Logistics 40:85–101
79. De P, Ghosh JB, Wells CE (1989) A note on the minimization of mean squared deviation of completion times about a common due date. Management Science 35:1143–1147
80. De P, Ghosh JB, Wells CE (1990) CON due-date determination and sequencing. Computers & Operations Research 17:333–342
81. De P, Ghosh JB, Wells CE (1990) Scheduling about a common due date with earliness and tardiness penalties. Computers & Operations Research 17:231–241
82. De P, Ghosh JB, Wells CE (1992) On the minimization of completion time variance with a bicriteria extension. Operations Research 40:1148–1155
83. De P, Ghosh JB, Wells CE (1993) On the general solution for a class of early/tardy problems. Computers & Operations Research 20:141–149
84. De P, Ghosh JB, Wells CE (1994) Solving a generalized model for CON due date assignment and sequencing. International Journal of Production Economics 34:179–185
85. De P, Ghosh JB, Wells CE (1994) Due-date assignment and early/tardy scheduling on identical parallel machines. Naval Research Logistics 41:17–32
86. Devillers R, Goossens J (2000) Liu and Layland's schedulability test revisited. Information processing letters 73:157–161
87. Dhamala TN, Kubiak W (2005) A brief survey of just-in-time sequencing for mixed-model systems. International Journal of Operations Research 2:38–47
88. Diamond JE, Cheng TCE (2000) Error bound for common due date assignment and job scheduling on parallel machines. IIE Transactions 32:445–448
89. Dileepan P (1993) Common due date scheduling problem with separate earliness and tardiness penalties. Computers & Operations Research 20:179–181
90. Ding FY, Cheng L (1993) A simple sequencing algorithm for mixed-model assembly lines in Just-in-Time production systems. Operations Research Letters 13:27–36
91. Ding FY, Cheng L (1993) An effective mixed-model assembly line sequencing heuristic for Just-in-Time production systems. Journal of Operations Management 11:45–50
92. Du J, Leung JYT, (1990) Minimizing total tardiness on one processor is NP-hard. Mathematics of Operations Research 15:483–495
93. Eilon S, Chowdhury IG (1976) Due dates in job shop scheduling. International Journal of Production Research 14:223–237
94. Eilon S, Chowdhury IG (1977) Minimizing waiting time variance in the single machine problem. Management Science 23:567–575

95. Emmons H (1987) Scheduling to a common due date on parallel uniform processors. Naval Research Logistics Quarterly 34:803–810
96. Feldman D, Biskup D Single-machine scheduling for minimizing earliness and tardiness penalties by meta-heuristic approach. Computers and Industrial Engineering 44:307–323
97. Finch BJ, Cox JF (1986) An examination of just-in-time management for small manufacturer: with an illustration. International Journal of Production Research 24:329–342
98. Fry TD, Leong K (1987) A bi-criterion approach to minimizing inventory costs on a single machine when early shipments are forbidden. Computers & Operations Research 14:363–368
99. Fry T, Armstrong R, Blackstone J (1987) Minimizing weighted absolute deviation in single machine scheduling. IIE Transactions 19:445–450
100. Fry TD, Leong K, Rakes T (1987) Single machine scheduling: a comparison of two solution procedures. OMEGA 15:277–282
101. Fry T, Armstrong R, Rosen LD (1990) Single machine scheduling to minimize mean absolute lateness. A heuristic solution. Computers & Operations Research 17:105–112
102. Fry T, Darby-Dowman K, Armstrong R (1996) Single machine scheduling to minimize mean absolute lateness. Computers & Operations Research 23:171–182
103. Garey MR, Johnson DS (1979) Computers and Intractability. WH Herman, San Francisco
104. Garey MR, Tarjan RE, Wilfong GT (1988) One-processor scheduling with symmetric earliness and tardiness penalties. Mathematics of Operations Research 13:330–348
105. Golhar DY, Stamm CL (1991) The just-in-time philosophy: A literature review International Journal of Production Research 29:657–676
106. Goldstein T, Miltenburg J (1988) The effects of pegging in the scheduling of Just-in-Time production systems. Working Paper #294, Faculty of Business, McMaster University, Hamilton
107. Gordon V, Proth J-M, Chu C (2002) Due date assignment and scheduling: SLK, TWK and other due date assignment models. Production Planning and Control 13:117–132
108. Gordon V, Proth J-M, Chu C (2002) A survey of the state-of-the-art of common due date assignment and scheduling research. European Journal of Operational Research 139:1–25
109. Groenevelt H (1993) The Just-in-Time System. In: Graves SC, Rinnooy Kan AHG, Zipkin PH (eds) Logistics of Production and Inventory, Handbooks in Operations Research and Management Science, Vol. 4, Elsevier, Amsterdam
110. Gupta JND, Lauff V, Werner F (2004) Two machine flow shop scheduling with nonregular criteria. Journal of Mathematical Modelling and Algorithms 3:123–151

111. Gupta MC, Gupta YP, Kumar A (1993) Minimizing flow time variance in a single machine system using genetic algorithms. European Journal of Operational Research 70:289–303
112. Gupta S, Sen T (1983) Minimizing a quadratic function of job lateness on a single machine. Engineering Costs and Production Economics 7:181–194
113. Gupta YP, Bector CR, Gupta MC (1990) Optimal schedule on a single machine using various due date determination methods. Computers in Industry 15:220–222
114. Halim AH, Miyazaki S, Ohia H (1994) Batch scheduling problems to minimize actual flow times of parts through the shop under JIT environment. European Journal of Operational Research 72:529–544
115. Hall NG (1986) Single and multiple processor models for minimizing completion variance. Naval Research Logistics Quarterly 33:49–54
116. Hall N, Kubiak W (1991) Proof of a conjecture of Schrage about the completion time variance problem. Operations Research Letters 10:467-472
117. Hall NG, Posner ME (1991) Earliness-tardiness scheduling problems, I: weighted deviation of completion times about a common due date. Operations Research 39:836–846
118. Hall NG, Kubiak W, Sethi SP (1991) Earliness-tardiness scheduling problems, II: deviation of completion times about a restrictive common due-date. Operations Research 39:847-856
119. Hao Q, Yang Z, Wang D, Li Z (1996) Common due date determination and sequencing using tabu search. Computers & Operations Research 23:409–417
120. Hardy GH, Littlewood JE, Polya G (1934) Inequalities. Cambridge University Press, New York
121. Hino CM, Ronconi DP, Mendes AB (2005) Minimizing earliness and tardiness penalties in a single-machine problem with a common due date. European Journal of Operational Research 160:190–201
122. Hiraishi K, Levner E, Vlach M Scheduling of parallel identical machines to maximize the weighted number of just-in-time jobs. Computers & Operations Research 29:841–848
123. Hoogeveen JA, van de Velde SL (1991) Scheduling around a small common due date. European Journal of Operational Research 55:237–242
124. Hoogeveen JA, Oosterhout H, van de Velde SL (1994) New lower and upper bounds for scheduling around a small common due date. Operations Research 42:102–110
125. Inman RR, Bulfin R (1991) Sequencing JIT mixed-model assembly lines. Management Science 37:901–904
126. Janiak A, Winczaszek M (2005) A single processor scheduling problem with a common due window assignment. In: Hein Fleuren, Dick den Hertog, Peter Kort (Eds) Operations Research. Springer Berlin, pp 213–220

127. Janiak A, Winczaszek M (2006) Common due window assignment in parallel processor scheduling problem with nonlinear penalty functions. Lecture Notes in Computer Science 3911:132–139
128. James RJW (1997) Using tabu search to solve the common due date early/tardy machine scheduling problem. Computers & Operations Research 24:199–208
129. James RJW, Buchanan TJ (1997) A neighbourhood scheme with a compressed solution space for the early/tardy scheduling problem. European Journal of Operational Research 102:513-527
130. Johnson DS (1974) Approximation algorithms for combinatorial problems. Journal of Computers and System Science 9:256–278
131. Józefowska J, Muras M (2003) Exact and heuristic approaches to single machine earliness-tardiness scheduling. Proc. 6th Internat. Conf. on Industrial Engineering and Production Management: 285–294
132. Józefowska J, Muras M (2004) A branch and bound algorithm for single machine earliness-tardiness scheduling. In: Oulamara A, Portmann MC (eds) Abstracts of the Ninth International Workshop on Project Management and Scheduling
133. Józefowska J, Muras M (2004) A Tabu Search Algorithm for Single Machine Just in Time Scheduling. In: Domek S., Kaszyński R. (Eds.), Proc. 10th IEEE Conference on Methods and Models in Automation and Robotics, pp 1255–1260.
134. Józefowska J, Józefowski Ł, Kubiak W (2006) Characterization of just in time sequencing via apportionment. In: Yan H, Yin G, Zhang Q (eds) Stochastic Processes, Optimization, and Control Theory Applications in Financial Engineering, Queueing Networks, and Manufacturing Systems/ A Volume in Honor of Suresh Sethi, Series: International Series in Operations Research & Management Science, Vol. 94, Springer Verlag
135. Józefowska J, Józefowski Ł, Kubiak W (2007) Apportionment methods and the Liu-Layland problem. submitted
136. Jurisch B, Kubiak W, Józefowska J (1997) Algorithms for MinClique scheduling problems. Discrete Applied Mathematics 72:115–139
137. Kahlbacher HG (1989) SWEAT - a program for a scheduling problem with earliness and tardiness penalties. European Journal of Operational Research 43:111–112
138. Kahlbacher HG (1993) Scheduling with monotonous earliness and tardiness penalties. European Journal of Operational Research 64:258–277
139. Kahlbacher HG, Cheng TCE (1995) Processing-plus-wait due dates in single machine scheduling. Journal of Optimization Theory and Applications 85:163–186
140. Kanet JJ (1981) Minimizing the average deviation of job completion times about a common due date. Naval Research Logistics Quarterly 28:643–651
141. Kanet JJ (1981) Minimizing variation of flow time in single machine systems. Management Science 27:1453–1459

142. Kanet JJ (1986) Tactically delayed versus non-delay scheduling: an experimental investigation. European Journal of Operational Research 24:99–105
143. Kanet JJ, Sridharan V (1991) PROGENITOR a genetic algorithm for production scheduling. Wirtschaftsinformatik 33:332–336
144. Kanet JJ, Sridharan SV (1998) The value of using scheduling information in planning material requirements. Decision Sciences 29:479-497
145. Kanet JJ, Sridharan V (2000) Scheduling with inserted idle time: problem taxonomy and literature review. Operations Research 48:99–110
146. Karacapilidis HG, Pappis CP (1993) Optimal due-date determination and sequencing of n jobs on a single machine using the SLK method. Computers in Industry 21:335–339
147. Karacapilidis HG, Pappis CP (1995) Form similarities of the CON and SLK due date determination methods. Journal of the Operational Research Society 46:762–770
148. Keyser TK, Sarper H (1991) A heuristic solution of the E/T problem with waiting cost and non-zero release times. Computers and Industrial Engineering 21:297-301
149. Kim YD, Yano CA (1994) Minimizing mean tardiness and earliness in single machine scheduling problems with unequal due dates. Naval Research Logistics 41:913–933
150. Koulamas C (1996) Single-machine scheduling with time windows and earliness/tardiness penalties. European Journal of Operational Research 91:190–202
151. Kovalyov M, Kubiak W (1999) A fully polynomial approximation scheme for the weighted earliness tardiness problem. Operations Research 47:757–761
152. Kovalyov M, Kubiak W, Yeomans JS (2001) A computational analysis of balanced JIT optimization algorithms. INFOR 39:299–316
153. Kubiak W (1993) Minimizing variation of production rates in just-in-time systems: A survey. European Journal of Operational Research 55:259-271
154. Kubiak W (1993) Completion time variance minimization on a single machine is difficult. Operations Research Letters 14:49–59
155. Kubiak W (1995) New results on the completion time variance minimization. Discrete Applied Mathematics 58:157–168
156. Kubiak W (2005) Solution of the Liu-Layland problem via bottleneck just-in-time sequencing. Journal of Scheduling 8:295–302.
157. Kubiak W (2003) Cyclic Just-In-Time Sequences Are Optimal. Journal of Global Optimization 27:333–347
158. Kubiak W (2003) Fair Sequences. Research Report, Faculty of Business Administration, Memeorial University of Newfoundland
159. Kubiak W (2003) On Small Deviations Conjecture. Bulletin of The Polish Academy of Science 51:189–203

160. Kubiak W (2005) Balancing Mixed-Model Supply Chains In: Avis D, Hertz A, Marcotte O (eds) Graph Theory and Combinatorial Optimization, GERAD 25th Anniversary Series Vol 8, Springer Science+Business Media Inc., New York
161. Kubiak W, Kovalyov M (1998) Product Rate Variation problem and greatest divisor property. Working Paper 98-15, Faculty of Business Administration, Memorial University of Newfoundland
162. Kubiak W, Sethi S (1991) A note on "Level schedules for mixed-model assembly lines in just-in-time production systems". Management Science 37:121–122
163. Kubiak W, Sethi S (1994) Optimal Just-in-Time Schedules for Flexible Transfer Lines. The International Journal of Flexible Manufacturing Systems 6:137–154
164. Kubiak W, Lou S, Sethi S (1990) Equivalence of mean flow time problems and mean absolute deviation problems. Operations Research Letters 9:371–374
165. Kubiak W, Steiner G, Yeomans JS (1997) Optimal Level Schedules for Mixed-Model Multi-Level Just-in-Time Assembly Systems. Annals of Operations Research 69:241–259
166. Lakshminarayan S, Lakshminarayan R, Papinou R, Rochette R (1978) Optimum single machine scheduling with earliness and tardiness penalties. Operations Research 26:1079–1082
167. Lawler A (1977) A pseudo-polynomial algorithm for sequencing jobs to minimize total tardiness. Annals of Discrete Mathematics 1:331–342
168. Lawler E, Moore J (1969) A functional equation and its applications to resource allocation and sequencing problems. Management Science 16:77–84
169. Lee CY, Choi JY (1995) A genetic algorithm for job sequencing problems with distinct due dates and general early tardy penalty weights. Computers & Operations Research 22:857–869
170. Lee CY, Danusaputro SL, Lin CS (1991) Minimizing weighted number of tardy jobs and weighted earliness-tardiness penalties about a common due date. Computers & Operations Research 18:379–389
171. Lee IS (1991) A worst case performance of the shortest-processing-time heuristic for single machine scheduling. Journal of the Operational Research Society 42:895–901
172. Leung JY-T (ed) (2004) Handbook of Scheduling. Algorithms, Models, and Performance Analysis. Chapman & Hall/CRC, Boca Raton
173. Li G (1997) Single machine earliness and tardiness scheduling. European Journal of Operational Research 96:546–558
174. Li CL, Cheng TCE (1994) The parallel machine min-max weighted absolute lateness scheduling problem. Naval Research Logistics 41:33–46
175. Liaw CF (1999) A branch-and-bound algorithm for the single machine earliness and tardiness scheduling problem. Computers & Operations Research 26:679–693

176. Liman SD, Panwalkar SS, Thongmee S (1998) Common due window size and location determination in a single machine scheduling problem. Journal of the Operational Research Society 49:1007–1010
177. Liu CL, Layland JW (1973) Scheduling algorithms for multiprogramming in a hard-real-time environment. Journal of the Association for Computing Machinery 20:46–61
178. Mahadev NVR, Pekec A, Roberts FS (1997) Effects of change of scale on optimality in a scheduling model with priorities and earliness/tardiness penalties. Mathematical and Computer Modelling 25:9–22
179. Mahadev NVR, Pekec A, Roberts FS (1998) On the Meaningfulness of Optimal Solutions to Scheduling Problems: Can an Optimal Solution be Non-Optimal? Operations Research 46:120–134
180. Manacher GK (1967) Production and stabilization of real-time task schedules. Journal of the Association for Computing Machinery 14:439–465
181. Manna DK, Prasad VR (1999) Bounds for the position of the smallest job in completion time variance minimization. European Journal of Operational Research 114:411–419
182. Mannur NR, Addagatla JB (1993) Heuristic algorithms for solving earliness-tardiness scheduling problems with machine vacations. Computers and Industrial Engineering 25:255–258
183. Mazzini R, Armentano VA (2001) A heuristic for single machine with early and tardy costs. European Journal of Operational Research 128:129–146
184. Meijer HG (1973) On a distribution problem in finite sets. Nederlandse Akademie van Wetenschappen Indagationes Mathematicae 35:9–17
185. Merten AG, Muller ME (1972) Variance minimization in single machine sequencing problem. Management Science 18:518–528
186. Mitlenburg J (1989) Level schedules for mixed-model assembly lines in just-in-time production systems. Management Science 35:192–207
187. Miltenburg J, Goldstein T (1991) Developing production schedules which balance part usage and smooth production loads for Just-in-Time production systems. Naval Research Logistics 38:893–910
188. Mitlenburg J, Sinnamon G (1989) Scheduling mixed-model multi-level just-in-time production systems. International Journal of Production Research 27:1487–1509
189. Miltenburg J, Steiner G, Yeomans S (1990) A dynamic programming algorithm for scheduling mixed-model Just-in-Time production systems. Mathematical and Computer Modelling 13:57–66
190. Mittenthal J, Raghavachari M, Rana AI (1995) V- and Λ-shaped properties for optimal single machine schedules for a class of non-separable penalty function. European Journal of Operational Research 86:262–269
191. Monden Y Toyota Production System. Institute of Industrial Engineering Press, Norcross, GA

192. Moreno N, Corominas A (2003) Solving the minsum productive rate variation problem (PRVP) as an assignment problem. In: 27 Congreso Nacional de Estadística e Investigación Operativa, Lleida
193. Multu O (1993) Comments on optimal due date determination and sequencing of n jobs on a single machine. Journal of the Operational Research Society 44:1062
194. Nandkeolyar U, Ahmed MU, Sundararaghavan PS (1993) Dynamic single machine weighted absolute deviation problem: predictive heuristics and evaluation. International Journal of Production Research 31:1453–1466
195. Nawaz M, Enscore EE, Ham I (1983) A heuristic algorithm for the m-machine, n-job flow-shop sequencing problem. The International Journal of Management Science 11:91–95
196. Ng CT, Cai X, Cheng TCE (1996) A tight lower bound for the completion time variance problem. European Journal of Operational Research 92:211–213
197. Ng CT, Cheng TCE, Bachman A, Janiak A (2002) Three scheduling problems with deteriorating jobs to minimize the total completion time. Information Processing Letters 81:327–333
198. Ng CT, Cheng TCE, Kovalyov M, Lam SS (2003) Single machine scheduling with a variable common due date and resource-dependent processing times. Computers & Operations Research 30:1173–1185
199. Nowicki E, Zdrzałka S (1990) A survey of results for sequencing problems with controllable processing times. Discrete Applied Mathematics 26:271–287
200. Oguz C, Dincer C (1994) Single machine earliness-tardiness scheduling problems using the equal-slack rule. Journal of the Operational Research Society 45:589–594
201. Ohno T (1988) Toyota production system: beyond large scale production. Productivity Press, New York
202. Ow PS, Morton TE (1988) Filtered beam search in scheduling. International Journal of Production Research 26:35–62
203. Ow PS, Morton TE (1989) The single machine early/tardy problem. Management Science 35:177-191
204. Panwalkar SS, Rajagopalan R (1992) Single machine sequencing with controllable processing times. European Journal of Operational Research 59:298–302
205. Panwalkar SS, Smith M, Seidmann A (1982) Common due date assignment to minimize total penalty for the one machine scheduling problem. Operations Research 30:391–399
206. Papadimitriou CH, Steiglitz K (1982) Combinatorial optimization. Algorithms and complexity. Prentice-Hall, Engelwood Cliffs
207. Potts CN, Van Wassenhoffe LN (1985) A branch and bound algorithm for the total weighted tardiness problem. Operations Research 33:363–377

208. Price W, Gravel M, Naskanda AL (1994) A review of optimisation models of Kanban-based production systems. European Journal of Operational Research 75:1–12
209. Quaddus M (1987) A generalized model of optimal due date assignment by linear programming. Journal of the Operational Research Society 38:353–359
210. Raghavachari M (1986) A V-shape property of optimal schedule of jobs about a common due date. European Journal of Operational Research 23:401–402
211. Raghavachari M (1988) Scheduling problems with non-regular penalty functions: a review. Opsearch 25:144–164
212. Roberts F (1995) A Functional Equation that Arises in Problems of Scheduling with Priorities and Lateness/Earliness Penalties. Mathematical and Computer Modelling 21:77–83
213. Schaller J (2004) Single machine scheduling with early and quadratic tardy penalties. Computers and Industrial Engineering 46:511–532
214. Schönhage A, Paterson M, Pippenger N (1976) Finding the median. Journal of Computer and System Sciences 13:184–199
215. Schrage L (1975) Minimizing the Time-in-System variance for a finite job set. Management Science 21:540–543
216. Seidmann A, Panwalkar S, Smith M (1981) Optimal assignment of due-dates for a single processor scheduling problem. International Journal of Production Research 19:393-399
217. Sen T, Gupta S (1984) A state-of-art survey of static scheduling research involving due dates. OMEGA 12:62-76
218. Sidney J (1977) Optimum single-machine scheduling with earliness and tardiness penalties. Operations Research 25:62–69
219. Sourd F, Kedad-Sidhoum S (2003) The one machine problem with earliness and tardiness penalties. Journal of Scheduling 6:533–549
220. Sridharan V, Zhou Z (1996) Dynamic non-preemptive single machine scheduling. Computers & Operations Research 23:1183-1190
221. Steiner G, Yeomans S (1993) Level schedules for mixed-model Just-in-Time processes. Management Science 39:728–735
222. Steiner G, Yeomans S (1994) A bicriterion objective for leveling the schedule of a mixed-model, JIT assembly processes. Mathematical & Computer Modelling 20:123–134
223. Steiner G, Yeomans S (1996) Optimal level schedules in mixed-model multi-level JIT assembly systems with pegging. European Journal of Operational Research 95:38–52
224. Still JW (1979) A class of new methods for Congressional Apportionment. SIAM Journal of Applied Mathematics 37:401–418
225. Sumichrast R, Russel R Evaluating mixed-model assembly line sequencing heuristics for Just-in-Time production systems. Journal of Operations Management 9:371–390

226. Sumichrast R, Russel R, Taylor B (1990) A comparative analysis of sequencing procedures for mixed-model assembly lines in Just-in-Time production systems. International Journal of Production Research 30:199–214
227. Sundararaghavan PS, Ahmed MU (1984) Minimizing the sum of absolute lateness in single-machine and multimachine scheduling. Naval Research Logistics Quarterly 31:325–333
228. Sung CS, Min JJ (2001) Scheduling in a two-machine flowshop with batch processing machine(s) for earliness/tardiness measure under a common due date. European Journal of Operational Research 131:95–106
229. Sung CS, Vlach M (2001) Just-in-time scheduling on parallel machines. The European Operational Research Conference, Rotterdam.
230. Szwarc W (1988) Minimizing absolute lateness in single machine scheduling with different due dates. Working Paper, University of Wisconsin, Milwaukee
231. Szwarc W (1989) Single machine scheduling to minimize absolute deviation of completion times from a common due date. Naval Research Logistics 36:663–673
232. Szwarc W (1990) Parametric precedence relations in single machine scheduling. Operations Research Letters 9:133–140
233. Szwarc W (1993) Adjacent ordering in single machine scheduling with earliness and tardiness penalties. Naval Research Logistics 40:229–243
234. Szwarc W (1996) The weighted common due date single machine scheduling problem revisited. Computers & Operations Research 23:255–262
235. Szwarc W, Mukhopadhyay SK (1995) Optimal timing schedules in earliness-tardiness single machine scheduling. Naval Research Logistics 42:1109–1114
236. Szwarc W, Mukhopadhyay SK (1996) Earliness and tardiness single machine scheduling with proportional weights. Journal of Global Optimization 9:1573–2916
237. Tarjan RE (1983) Data structures and networks algorithms. Society for Industrial and Applied Mathematics, Philadelphia
238. Tijdeman R (1980) The chairman assignment problem. Discrete Mathematics 32:323–330
239. van den Akker M, Hoogeven H, van de Velde S (1999) Parallel machine scheduling by column generation. Operations Research 47:862–872
240. van de Velde S (1990) A simpler and faster algorithm for optimal total-work-content-power due date determination. Mathematical and Computer Modelling 13:81–83
241. Vanhoucke M, Demeulemeester E, Herroelen W An Exact Procedure for the Resource-Constrained Weighted EarlinessŰTardiness Project Scheduling Problem. Annals of Operations research 102:179–196

242. Vani V, Raghavachari M (1987) Deterministic and random single machine sequencing with variance minimization. Operations Research 35:111–120
243. Ventura JA, Radhakrishnan S (2003) Single machine scheduling with symmetric earliness and tardiness penalties. European Journal of Operational Research 144:598–612
244. Ventura JA, Weng MX (1995) An improved dynamic programming algorithm for the single machine mean absolute deviation problem with a restrictive common due date. Operations Research Letters 17:149–152
245. Ventura JA, Kim D, Garriga F (2002) Single machine earliness-tardiness scheduling with resource-dependent release dates. European Journal of Operational Research 142:52–69
246. Viswanathkumar G, Srinnivasan G (2003) A branch and bound algorithm to minimize completion time variance on a single processor. Computers & Operations Research 30:1135–1150
247. Vollman TE, Berry WL, Wybark DC (1997) Manufacturing Planning and Control Systems, 4th ed. McGraw-Hill, New York
248. Wan G, Benjamin P-CY (2002) Tabu search for single machine scheduling with distinct due windows and weighted earliness/tardiness penalties. European Journal of Operational Research 142:271–281
249. Weeks JK (1979) A simulation study of predictable due-dates. Management Science 25:363–373
250. Weeks JK, Fryer JS (1977) A methodology for assigning minimum cost due-dates. Management Science 23:872–881
251. Weng X, Ventura A (1996) Scheduling around a small common due date to minimize mean squared deviation of completion times. European Journal of Operational Research 88:328–335
252. Yano CA, Kim YD (1986) Algorithms for single machine scheduling problems minimizing tardiness and earliness. Technical Report, University of Michigan, Ann Arbor
253. Yano CA, Kim YD (1991) Algorithms for a class of single-machine weighted tardiness and earliness problem. European Journal of Operational Research 52:167–178, European Journal of Operational Research, 81:663–664
254. Yeung WK, Oguz C, Cheng TCE (2001) Minimizing weighted number of early and tardy jobs with a common due window involving location penalty. Annals of Operations Research 108:33–54
255. Yeung WK, Oguz C, Cheng TCE (2001) Single-machine scheduling with a common due window. Computers & Operations Research 28:157–175
256. Yeung WK, Oguz C, Cheng TCE (2004) Two-stage flowshop earliness and tardiness machine scheduling involving a common due window. International Journal of Production Economics 90:421–434
257. Yoo WS, Martin-Vega LA (2001) Scheduling single-machine problems for on-time delivery. Computers & Industrial Engineering 39:371–392

258. Zdrzałka S (1991) Scheduling jobs on a single machine with release dates, delivery times and controllable processing times: worst case analysis. Operational Research Letters 10:519–524
259. Zionts S (1974) Linear and integer programming. Prentice-Hall, Englewood Hills

Index

Early Titles in the
INTERNATIONAL SERIES IN OPERATIONS RESEARCH & MANAGEMENT SCIENCE
Frederick S. Hillier, Series Editor, *Stanford University*

Saigal/ *A MODERN APPROACH TO LINEAR PROGRAMMING*

Nagurney/ *PROJECTED DYNAMICAL SYSTEMS & VARIATIONAL INEQUALITIES WITH APPLICATIONS*

Padberg & Rijal/ *LOCATION, SCHEDULING, DESIGN AND INTEGER PROGRAMMING*

Vanderbei/ *LINEAR PROGRAMMING*

Jaiswal/ *MILITARY OPERATIONS RESEARCH*

Gal & Greenberg/ *ADVANCES IN SENSITIVITY ANALYSIS & PARAMETRIC PROGRAMMING*

Prabhu/ *FOUNDATIONS OF QUEUEING THEORY*

Fang, Rajasekera & Tsao/ *ENTROPY OPTIMIZATION & MATHEMATICAL PROGRAMMING*

Yu/ *OR IN THE AIRLINE INDUSTRY*

Ho & Tang/ *PRODUCT VARIETY MANAGEMENT*

El-Taha & Stidham/ *SAMPLE-PATH ANALYSIS OF QUEUEING SYSTEMS*

Miettinen/ *NONLINEAR MULTIOBJECTIVE OPTIMIZATION*

Chao & Huntington/ *DESIGNING COMPETITIVE ELECTRICITY MARKETS*

Weglarz/ *PROJECT SCHEDULING: RECENT TRENDS & RESULTS*

Sahin & Polatoglu/ *QUALITY, WARRANTY AND PREVENTIVE MAINTENANCE*

Tavares/ *ADVANCES MODELS FOR PROJECT MANAGEMENT*

Tayur, Ganeshan & Magazine/ *QUANTITATIVE MODELS FOR SUPPLY CHAIN MANAGEMENT*

Weyant, J./ *ENERGY AND ENVIRONMENTAL POLICY MODELING*

Shanthikumar, J.G. & Sumita, U./ *APPLIED PROBABILITY AND STOCHASTIC PROCESSES*

Liu, B. & Esogbue, A.O./ *DECISION CRITERIA AND OPTIMAL INVENTORY PROCESSES*

Gal, T., Stewart, T.J., Hanne, T./ *MULTICRITERIA DECISION MAKING: Advances in MCDM Models, Algorithms, Theory, and Applications*

Fox, B.L./ *STRATEGIES FOR QUASI-MONTE CARLO*

Hall, R.W./ *HANDBOOK OF TRANSPORTATION SCIENCE*

Grassman, W.K./ *COMPUTATIONAL PROBABILITY*

Pomerol, J-C. & Barba-Romero, S./ *MULTICRITERION DECISION IN MANAGEMENT*

Axsater, S./ *INVENTORY CONTROL*

Wolkowicz, H., Saigal, R., & Vandenberghe, L./ *HANDBOOK OF SEMI-DEFINITE PROGRAMMING: Theory, Algorithms, and Applications*

Hobbs, B.F. & Meier, P./ *ENERGY DECISIONS AND THE ENVIRONMENT: A Guide to the Use of Multicriteria Methods*

Dar-El, E./ *HUMAN LEARNING: From Learning Curves to Learning Organizations*

Armstrong, J.S./ *PRINCIPLES OF FORECASTING: A Handbook for Researchers and Practitioners*

Balsamo, S., Personé, V., & Onvural, R./ *ANALYSIS OF QUEUEING NETWORKS WITH BLOCKING*

Bouyssou, D. et al./ *EVALUATION AND DECISION MODELS: A Critical Perspective*

Hanne, T./ *INTELLIGENT STRATEGIES FOR META MULTIPLE CRITERIA DECISION MAKING*

Saaty, T. & Vargas, L./ *MODELS, METHODS, CONCEPTS and APPLICATIONS OF THE ANALYTIC HIERARCHY PROCESS*

Chatterjee, K. & Samuelson, W./ *GAME THEORY AND BUSINESS APPLICATIONS*

Hobbs, B. et al./ *THE NEXT GENERATION OF ELECTRIC POWER UNIT COMMITMENT MODELS*

Vanderbei, R.J./ *LINEAR PROGRAMMING: Foundations and Extensions, 2nd Ed.*

Kimms, A./ *MATHEMATICAL PROGRAMMING AND FINANCIAL OBJECTIVES FOR SCHEDULING PROJECTS*

Baptiste, P., Le Pape, C. & Nuijten, W./ *CONSTRAINT-BASED SCHEDULING*

Feinberg, E. & Shwartz, A./ *HANDBOOK OF MARKOV DECISION PROCESSES: Methods and Applications*

Ramík, J. & Vlach, M./ *GENERALIZED CONCAVITY IN FUZZY OPTIMIZATION AND DECISION ANALYSIS*

Song, J. & Yao, D./ *SUPPLY CHAIN STRUCTURES: Coordination, Information and Optimization*

Kozan, E. & Ohuchi, A./ *OPERATIONS RESEARCH/ MANAGEMENT SCIENCE AT WORK*

Bouyssou et al./ *AIDING DECISIONS WITH MULTIPLE CRITERIA: Essays in Honor of Bernard Roy*

Cox, Louis Anthony, Jr./ *RISK ANALYSIS: Foundations, Models and Methods*

Dror, M., L'Ecuyer, P. & Szidarovszky, F./ *MODELING UNCERTAINTY: An Examination of Stochastic Theory, Methods, and Applications*

Dokuchaev, N./ *DYNAMIC PORTFOLIO STRATEGIES: Quantitative Methods and Empirical Rules for Incomplete Information*

Sarker, R., Mohammadian, M. & Yao, X./ *EVOLUTIONARY OPTIMIZATION*

Demeulemeester, R. & Herroelen, W./ *PROJECT SCHEDULING: A Research Handbook*

Gazis, D.C./ *TRAFFIC THEORY*

Zhu/ *QUANTITATIVE MODELS FOR PERFORMANCE EVALUATION AND BENCHMARKING*

Ehrgott & Gandibleux/ *MULTIPLE CRITERIA OPTIMIZATION: State of the Art Annotated Bibliographical Surveys*

Bienstock/ *Potential Function Methods for Approx. Solving Linear Programming Problems*

Matsatsinis & Siskos/ *INTELLIGENT SUPPORT SYSTEMS FOR MARKETING DECISIONS*

Alpern & Gal/ *THE THEORY OF SEARCH GAMES AND RENDEZVOUS*

Hall/ *HANDBOOK OF TRANSPORTATION SCIENCE - 2nd Ed.*

Glover & Kochenberger/ *HANDBOOK OF METAHEURISTICS*

Graves & Ringuest/ *MODELS AND METHODS FOR PROJECT SELECTION: Concepts from Management Science, Finance and Information Technology*

Hassin & Haviv/ *TO QUEUE OR NOT TO QUEUE: Equilibrium Behavior in Queueing Systems*

Gershwin et al/ *ANALYSIS & MODELING OF MANUFACTURING SYSTEMS*

Maros/ *COMPUTATIONAL TECHNIQUES OF THE SIMPLEX METHOD*

Harrison, Lee & Neale/ *THE PRACTICE OF SUPPLY CHAIN MANAGEMENT: Where Theory and Application Converge*

Shanthikumar, Yao & Zijm/ *STOCHASTIC MODELING AND OPTIMIZATION OF MANUFACTURING SYSTEMS AND SUPPLY CHAINS*

Nabrzyski, Schopf & Węglarz/ *GRID RESOURCE MANAGEMENT: State of the Art and Future Trends*

Thissen & Herder/ *CRITICAL INFRASTRUCTURES: State of the Art in Research and Application*

Carlsson, Fedrizzi, & Fullér/ *FUZZY LOGIC IN MANAGEMENT*

Soyer, Mazzuchi & Singpurwalla/ *MATHEMATICAL RELIABILITY: An Expository Perspective*

Chakravarty & Eliashberg/ *MANAGING BUSINESS INTERFACES: Marketing, Engineering, and Manufacturing Perspectives*

Talluri & van Ryzin/ *THE THEORY AND PRACTICE OF REVENUE MANAGEMENT*

Kavadias & Loch/ *PROJECT SELECTION UNDER UNCERTAINTY: Dynamically Allocating Resources to Maximize Value*

Brandeau, Sainfort & Pierskalla/ *OPERATIONS RESEARCH AND HEALTH CARE: A Handbook of Methods and Applications*

Cooper, Seiford & Zhu/ *HANDBOOK OF DATA ENVELOPMENT ANALYSIS: Models and Methods*

Luenberger/ *LINEAR AND NONLINEAR PROGRAMMING, 2nd Ed.*

Early Titles in the
INTERNATIONAL SERIES IN OPERATIONS RESEARCH & MANAGEMENT SCIENCE
(Continued)

Sherbrooke/ *OPTIMAL INVENTORY MODELING OF SYSTEMS: Multi-Echelon Techniques, Second Edition*

Chu, Leung, Hui & Cheung/ *4th PARTY CYBER LOGISTICS FOR AIR CARGO*

Simchi-Levi, Wu & Shen/ *HANDBOOK OF QUANTITATIVE SUPPLY CHAIN ANALYSIS: Modeling in the E-Business Era*

***** A list of the more recent publications in the series is at the front of the book *****